ATF Science and Theology Series: 2

Interdisciplinary Perspectives on Cosmology and Biological Evolution

edited by Hilary D Regan and Mark Wm Worthing

The ATF Science and Theology Series is a publication of the Australian Theological Forum. Each volume is a collection of essays by one or a number of authors in the area of science and theology. The Series addresses particular themes in the nexus between the two disciplines, and draws upon the expertise of both scientists and theologians.

ATF Science and Theology Series
Series Editor: Mark Wm Worthing

1: *God, Life, Intelligence and the Universe*, edited by Terence J Kelly SJ and Hilary D Regan

Interdisciplinary Perspectives on Cosmology and Biological Evolution

edited by

Hilary D Regan
&
Mark Wm Worthing

Australian Theological Forum
Adelaide

Text copyright © 2002 ATF for all papers in this volume.

All rights reserved. Except for any fair dealing permitted under the Copyright Act, no part of this book may be reproduced by any means without prior permission. Inquiries should be made to the publisher.

First published January 2002

National Library of Australia
Cataloguing-in-Publication data

Interdisciplinary perspectives on cosmology and evolutionary biology

Bibliography
ISBN 0 9586399 9X

1. Cosmology 2. Religion and science I. Regan, Hilary D. II. Worthing, Mark William. III. Australian Theological Forum (Series: ATF science and theology series; 2).

231.765

Published by
Australian Theological Forum
P O Box 504
Hindmarsh
SA 5007
Australia

ABN: 68 314 074 034
www.atf.org.au

Printed by Openbook Publishers, Adelaide, Australia

Contents

Contributors   vii
Introduction   ix
    Hilary D Regan

Part One
Historical Perspectives

1. The Changing Relations between Science and Religion   3
   John Hedley Brooke
2. 'God's Two Books': Revelation, Theology and Natural Science in the Christian West   19
   Peter J Hess

Part Two
Philosophical Perspectives

3. Why Christians Should Be Physicalists   52
   Nancey Murphy
4. How Physicalists Can Avoid Being Reductionists   69
   Nancey Murphy
5. Response to Nancey Murphy   91
   Denis Edwards
6. Reflections on Artificial Intelligence, Emergence and Agency   99
   Adrian Wyard

Part Three
Theological Perspectives

7. Science, the Laws of Nature and Divine Action   117
   William R Stoeger SJ
8. Cosmology and a Theology of Creation   128
   William R Stoeger SJ

9.	Response to William R Stoeger SJ *Denis Edwards*	146
10.	God, Process and Cosmos: Is God Simply Going Along for the Ride? *Mark Wm Worthing*	153
11.	Evolution and the Christian God *Denis Edwards*	172
12.	Historiographical Resources for Teaching Religion and Science	195
13.	CTNS Brief Bibliography in Science and Religion	199
14.	Author Index	210
15.	Subject Index	215

Contributors

John Hedley Brooke is Andreas Idreos Professor of Science and Religion, Oxford University, Oxford.

Peter MJ Hess is Associate Program Director, Science and Religion Course Program, Berkeley, California.

Nancey Murphy is Professor of Christian Philosophy at Fuller Theological Seminary, Pasadena, California, and board member of the Center for Theology and the Natural Sciences, Berkeley, California.

Denis Edwards teaches theology at Catholic Theological College, at the Adelaide College of Divinity, and is a member of the board of the Flinders University/Adelaide College of Divinity Centre for Theology, Science and Culture.

Adrian Wyard is Founder and Director of the Counter Balance Foundation, in Seattle, Oregon, and board member of the Center for Theology and the Natural Sciences, Berkeley, California.

William R Stoeger SJ is a staff astrophysicist at the Vatican Observatory Research Group, Tuscon Arizona, and Adjunct Associate Professor of Astronomy, University of Arizona.

Mark Wm Worthing is Dean of Studies, Tabor College, Adelaide, South Australia, and Regional Director, Australia and New Zealand, for the Science and Religion Course Program.

Hilary D Regan is Secretary of the Australian Theological Forum and Extension Consultant with the Science and Religion Course Program, Berkeley, California.

Introduction

Hilary D Regan

Biological and cosmological evolution have been major areas of debate in both theology and the natural sciences for much of the last century. Therefore it is appropriate that there is a book at the beginning of a new century which brings together scientists, theologians, philosophers and historians to examine the current interplay in these complex fields.

This book is concerned with heuristics, the knowledge which may arise in the nexus, or the overlapping, of the interests of both those coming from the perspective of the wide range of disciplines in the scientific world and those from a theological perspective. It is built on the assumption that when both groups work together, responding out of mutual concern for truth, breaking down the historical barriers that have developed and which have negated and discouraged a dialogue, and attempting to find new ways of interacting, this will be of benefit to both disciplines, and, ultimately, all of humanity.

Likewise, the book suggests that science and theology are both at their best when they not only understand each other but as a result of this mutual understanding work together in addressing the issues of the day. Both science and theology need to pursue their own agendas, and yet can be and are interconnected in their pursuit of truth. Neither is forced to choose between isolation or assimilation. There is a prophetic role for both science and theology in a world in need of transformation, and that through such cooperation this transformation can be generated. Suspicion and fear can be based on a lack of knowledge, and breaking down misunderstandings through the building of bridges is achieved

Hilary D Regan

through various forms of dialogue. This book is a heurisitical tool of 'prophetic' value for a 'prophetic' task.

The collection of essays contained in this volume comes from the first Workshop to be held in Australia sponsored by the Center for Theology and the Natural Sciences (CTNS) Science and Religion Course Program (SRCP), and funded by the John Templeton Foundation.

The book is published by the Australian Theological Forum (ATF), an independent Australian based ecumenical and interdisciplinary research and publishing body. The Forum seeks to network with like minded bodies both within Australia and internationally, and so cooperation with CTNS fits well with its charter. The editors acknowledge the assistance given by the Program for the production of this volume and others in the ATF's Science and Theology Series, which is but one of a number of Series in the Forum's publishing program.

Situating this volume of essays in the Asia Pacific region is important. The importance is derived from its social and political place in international and regional relationships. The meeting was held in late January 2001, the first month of the new millennium. Just over one hundred academics, from various scientific and non-scientific backgrounds—historians, philosophers, theologians—seminary professors, ministers of religion, as well as interested lay-folk, gathered together in Adelaide, South Australia, for five days. Some would ultimately go on to receive the SRCP Course Award and teach a tertiary-based course in Science and Religion and some of these may use this volume as a reference book in a course being taught. The participants came from all over the globe, with all continents represented, to listen to and engage with each other and with speakers who covered the fields of history, philosophy, theology and science, while examining cosmology and evolutionary biology.

On the international level, George W Bush had recently been elected as the new American President, peace talks in both the Middle East and in Northern Ireland had been faltering for

Introduction

some time and various attempts to have them restarted or moved on had been stalled. International terrorism of various forms was escalating, later that year culminating on 11 September 2001 with the deaths of thousands in the USA. Within the region of Asia-Pacific, there was a steady increase in the number of refugees seeking a better life outside of their troubled countries with a response of nationalistic closed-door policies by some governments to protect 'national sovereignty'. At the same time international peace-keeping forces were seeking to maintain the peace and facilitate the rebuilding of the broken nation of East Timor, as well as other nations around the globe.

In the scientific world the task of mapping the human genome continued all over the world with the release of the whole genome in the middle of 2001. With this was the hope of not only understanding more about the human person, but also the possibility of developing a response to many diseases and genetic disorders. Space exploration continued with the development of the international space station, while at the same time the whole world is cognisant of the statistics of the numbers who die of malnutrition and poverty each day and the costs needed to prevent these unnecessary deaths.

This book of essays, while not addressing any of these issues directly, does so indirectly by addressing the nature of divine action and divine power in creation revelation, human freedom, the laws of nature, cosmology, evolutionary biology, the role of artificial intelligence and information technology, and leads into a discussion of Christian theism, Christology, and pneumatology. In addressing these topics it does not seek specific solutions or answers to the mysteries of the world. It aims to raise perspectives for how one might interpret the world in need of a scientific and theological response, and of how in all the issues, as well as those of time, eternity and theodicy, there is the eschatological promise of a new creation in Christ. As Dr Denis Edwards writes in this volume, it means 'living in constant expectation of God as the absolute future', by

Hilary D Regan

following a biblically based theology which teaches us 'to value and love the present moment as the gift of grace'. To this end, we entrust this book to all seeking a better world, whether from the perspective of the academy, scientific or religious, the church or any other audience.

In terms of the structure of the volume, it is divided into three sections which give historical, philosophical and theological perspectives on biological and cosmological evolution. Continuing the objective of facilitating and promoting dialogue, some papers have a response while others do not. The responses aim to tease out the underlying presuppositions of the papers and push the conversation further. This discussion, this heuristic, continues in any place where this book of essays is read and debated.

January 2002

Part One

Historical Perspectives

The Changing Relations between Science and Religion

John Hedley Brooke

One of the pleasing features of the explosion of interest in the field of 'science and religion' has been the growing recognition that it is more helpful to speak of sciences and religions than to pursue the chimerical goal of defining 'science' and 'religion', both in the singular, with a view to giving a definitive account of their mutual relations or relevance. The sciences of quantum mechanics and evolutionary biology might be correlated with religious concerns in quite different ways, and those concerns might vary considerably from one religion to another. There is, however, another challenge to essentialism, arising not so much from the fact that sciences as well as religions have to be differentiated, but from the fact that the relations between them have changed over time. It is this second challenge with which I shall be most concerned in this essay. My aim is not so much to describe the changed relations in detail as to explore some of the implications of the fact that they do change. It is wise to recognise that 'science' is not a given, but a word that masks a range of diversified practices that have themselves evolved. In their practices the alchemists of the Renaissance were doing something very different from their contemporary astronomers, and very different again from subsequent mechanical philosophers, such as Descartes, who sought a mechanistic account of how God had made the world.

Even the most superficial acquaintance with the history of Western philosophy would indicate the fact that sweeping changes have occurred.

John Hedley Brooke

In antiquity, as David Lindberg has argued, 'there was nothing . . . corresponding to modern science as a whole or to such branches of modern science as physics, chemistry, geology, zoology, and psychology'. Rather, the world of the intellect had a unity it does not have today: 'It was not sharply divisible into separate disciplines, such as metaphysics, theology, epistemology, ethics, natural science, and mathematics, but presented itself as a relatively unified and coherent whole'. During the medieval period, theology itself was deemed 'queen of the sciences' where the word 'science' denoted an organised body of knowledge. During the seventeenth century there were many innovations in the study of natural philosophy, but it was possible for John Locke to say at the end of that century that he could not see how natural philosophy could become a science—in the Aristotelian sense of providing certain knowledge, deductively demonstrated. Locke's contemporary Isaac Newton unequivocally affirmed that discussion of God was part of what it meant to engage in natural philosophy. As late as the 1830s when he first coined the word 'scientist', it was possible for the Cambridge philosopher William Whewell to argue that the study of the life sciences not merely included, but was regulated by, the idea of final causes. Examples such as these not only provide arguments against an essentialist approach but also remind us that the so-called 'relations between science and religion' do not exist in some kind of cyberspace. It is we who construct them.

This is one of many lessons we can learn from historical investigation. Another is that history itself is often deployed to support partisan interests, both in favour of and against secular narratives. One of the best known examples would be John William Draper's *History of the Conflict between Religion and Science* (1874), in which an inherent conflict between science and the Roman Catholic Church provided a topical theme when papal infallibility on matters of faith and doctrine was being strongly affirmed. So embedded was Draper's essentialism that he was able to persuade himself that the church had

The Changing Relations between Science and Religion

condemned Galileo whilst knowing he was *right*. For Draper there had to be conflict because it was in the nature of the sciences to change, whereas it was in the nature of religion, particularly Catholic Christianity, to remain static. Essentialist arguments can of course take other, very different forms, as when religious apologists sometimes maintain an underlying harmony between 'science' and 'religion' when both are properly understood.

Historians are apt to feel uncomfortable when confronted with such arguments because they are keenly aware of changes in both scientific and religious sensibility that make the meta-level claims difficult to sustain. In particular (and this is what makes the subject so enthralling), new forms of science have been perceived to have implications for theology and new forms of religious practice for the sciences. As an example of the latter, Peter Harrison has argued that a new style of Protestant hermeneutics in the sixteenth and seventeenth centuries had implications for how the book of nature was read. To the gradual abandonment of allegorical and typological readings of scriptural texts, in favour of a single, literal interpretation, there corresponded the gradual displacement of emblematic readings of natural objects by a study of their practical utility and the causal relations between them. Newton would personify this quest for a single definitive meaning for each biblical text and a single correct explanation (usually his own!) for each natural phenomenon.

Conversely, new forms of science have often been appropriated for religious purposes, giving the lie to our routine images of religions in retreat as the sciences advance. Newton's science, especially his gravitational theory, was seized by his contemporary William Whiston as proof against the overly mechanistic science of Descartes which, in his view (and Newton's too) had left too little room for an active Providence. The invisible, 'immechanical' forces that Whiston saw in Newton's science provided ammunition for theology's engagement with the atheist who might doubt the reality of an invisible, spiritual world. There are many other such examples that show how scientific progress has not always come to the aid of the secularist. In nineteenth century France, Louis Pasteur's dismissal of the spontaneous generation of microscopic life forms could be hailed as an argument against materialism. One of the lessons we learn from the fact that the relations between scientific and religious beliefs have changed is that we should not compress the changes into a streamlined process of secularisation.

There are a least two other preliminary lessons that emerge from the analysis of change in the content of the sciences. Both concern apologetic strategies that have proved attractive to religious thinkers but which on closer inspection have also proved damaging. The temptation to make theological claims on the basis of that which is currently inexplicable in scientific terms can lead to a god-of-the-gaps who becomes progressively more redundant as the sciences expand their horizons. Such a god is too small in the first place and shrinks even further as the sciences advance. In this context Newton is often cited for his contention that the solar system would eventually collapse as a consequence of the retarding effects of an aether on planetary motions. His insistence that a 'reformation' of the system would be required from time to time was associated with arguments for a divine initiative in ensuring that stability would be restored. As it happens Newton did not rule out a role for secondary causes in effecting the divine will; but it was easy for Leibniz to caricature his god as a clockmaker who had made a defective clock. When Laplace later showed that the solar system could stabilise itself, Newton's appeal to Providence could appear misguided. An apologia based on a god-of-the gaps plays into the hands of aggressively anti-religious scientists who draw the conclusion that all they need to do to bring religious traditions into disrepute is to show that it is possible, in principle, to account for natural phenomena in naturalistic terms. The mistake has been repeated many times, as when the American botanist Asa Gray tried to persuade

The Changing Relations between Science and Religion

Charles Darwin that if he could not explain the origin of the variation on which natural selection worked he should ascribe its source to a deity in control of the evolutionary process. Gray undoubtedly believed that, whatever the mechanism, the ultimate source lay in the divine will, but his advice to Darwin ran the obvious risk of obsolescence as the origins of variation were brought with the purview of a science of genetics. One may only recognise the mistake in retrospect, but that is precisely why the study of historical change can be instructive.

A second apologetic strategy vulnerable to scientific change was recognised in antiquity by Augustine. This concerns the exegetical practice of making the interpretation of sacred texts dependent on the latest developments in secular or scientific learning. Augustine's concern was that to tie the meaning of Scripture to fashionable but ephemeral ideas would be deeply embarrassing once those ideas were displaced. A similar point was made by James Clerk Maxwell in the nineteenth century: to construct a theology on the basis of current scientific theory was inadvisable because of the dynamics of scientific change. It is a lesson that was often learnt the hard way. Clerical geologists in England during the early years of the nineteenth century sometimes claimed, as did William Buckland in Oxford, that geology provided independent corroboration of a universal flood, ostensibly confirming the biblical story of the deluge. But this intimate interdependence had to be sacrificed when alternative scientific accounts were proposed for those features of the earth's surface, such as U-shaped valleys, that had been ascribed to rushing water. Both Buckland and his opposite number in Cambridge, Adam Sedgwick, publicly acknowledged their mistake. It is not difficult to understand the alacrity with which favourable scientific disclosures may be seized for strategic purposes; but such examples warn against the overburdening of the sciences with religious meaning.

I have focused so far on the implications of scientific change for the appraisal of apologetic strategies. But the study of religious change, especially in the life of an individual, can also be illuminating. It may be tempting to ask what religious beliefs were held by the scientists of the past, and one is often surprised by the answers. But there is an ulterior problem in that those beliefs often change during a lifetime and it can be unhelpful to pigeon-hole them. Charles Darwin showed that you cannot pigeon-hole pigeons, so plastic were they in the hands of the breeders. It would be surprising if he and his religious beliefs could be easily categorised. There is more than one issue here. There could be fluctuation of belief within a short time frame. Darwin said himself that his beliefs often fluctuated, sometimes more strongly theistic, sometimes more accurately agnostic. But he also added that, as he grew older, agnosticism had become his more prevalent position, indicating that there could also be significant change over longer periods. It is a simple point but nevertheless important. His Christianity had first given way to a kind of deism. The Darwin who had once trained for the Anglican ministry later found that fact almost 'ludicrous'. The most sensitive scholarship on Darwin's loss of faith testifies to another important consideration—that the erosion of religious belief need not be due primarily to the potency of new forms of science. Darwin's own renunciation of Christianity had much to do with the moral repugnance he felt towards the doctrine of eternal damnation, but his most affecting existential crisis occurred with the death of his cherished daughter Annie at the tender age of ten. It is possible to overplay the role of science in the processes of secularisation.

This last point deserves a moment's reflection because the spiritual trajectories of scientific thinkers have not always gone from faith to doubt. One of the most prolific popularisers of geology in the nineteenth century, the Scottish Evangelical Hugh Miller, became fascinated by fossil forms before he was brought to faith in a redeemer God. But once drawn into the faith he began to see his fossils in a new light. In the beauty of an ammonite he saw structures that had presaged, and in other forms surpassed, the glories of human architecture in our Gothic cathedrals. Humans had not necessarily imitated nature

The Changing Relations between Science and Religion

consciously but there was such a comparable aesthetic sensibility in the human and the divine mind, that Miller found in geology an endorsement of the doctrine of *imago dei*. We should not assume that advances in geology were invariably destructive of a conservative creed. Later in the nineteenth century, Darwin's disciple in the science of animal behaviour George Romanes would vacillate between agnosticism and theism but with the theism eventually proving the more resilient.

In much contemporary literature on science and religion, the emphasis falls on reinterpreting religious language in the light of scientific knowledge. That trend is itself a reflection of one of the most significant changes that has occurred in the history of Western culture. In earlier periods the greater authority was vested in religious institutions and traditions, which meant that new forms of science were often judged according to their theological implications. New sciences had to create new spaces for themselves that could be legitimated in theological terms. A profound movement away from this situation is usually traced to the eighteenth century, when in various parts of Europe an explosion occurred in more secular forms of learning. In some cases, as with Voltaire's popularisation of Newton, the sciences were explicitly turned against the authority of the Catholic Church. The transition was of course very gradual and varied with location. The very attempt to give scientific support to biblical authority, in Newton or Whiston for example, might be seen as a symptom of the incipient change. Was the authenticity of revelation so much in doubt by the end of the seventeenth century that it needed such support? Whatever the answer, the fact that faith was gradually transferred from confidence in revelation to confidence in scientific methods must not blind us to the fact that the relations between the two have not always been construed in one direction only. There has been two-way traffic with theological ideas and debates contributing to the milieu in which scientific ideas are formulated and assimilated. The process of dialogue has not always been one-way.

John Hedley Brooke

I have already referred to Peter Harrison's thesis that new ways of reading the Bible during the period of Reformation and Counter Reformation helped to prepare for new readings of the book of nature. The theological squabbles of the period with their political ramifications also had an effect, in that the Bible was sometimes treated as a set of proof texts to be used in defence of particular concepts of authority. This was one source of tension in the Galileo affair. Galileo himself wanted to argue that biblical texts had been accommodated to the minds of ordinary people and were therefore not to be construed as determinants of truth on physical matters. But he had to face the oppressive caution of Cardinal Bellarmine who, in a personal gloss on the ruling of the Council of Trent, effectively made the stasis of the earth a matter of faith and morals over which the Church had jurisdiction. At times of crisis in matters of religious authority, the sciences may take on a complexion that might not otherwise have been so strongly coloured. A comparable situation arose in nineteenth century England when Darwinian evolution was made public at a time of crisis for the Anglican Church, struggling with the desertion of intellectuals and with the implications for biblical authority of the new forms of historical criticism. Put another way, one might say that new forms of biblical criticism helped pave the way for the assimilation of evolutionary theory among those whose belief in Providence was not grounded in a literal reading of Scripture. In earlier periods it is even possible that theological doctrines provided a crucial resource for scientific activism. Francis Bacon, in his diplomacy for the applied sciences, had exploited the doctrine of the Fall to legitimate altruistic application of new empirical knowledge. The justification lay in the promise of restoration—of restoring humankind's dominion over nature that had been God's intention for humanity but which had been sacrificed through the primal sin. Dreams of a science-based utopia have sometimes been the secular equivalent of millenarian expectation. It was even possible for Bacon to argue that practice of the sciences would foster the Christian virtue of

The Changing Relations between Science and Religion

humility because empirical methods threatened the arrogance of those whose certainties were derived from books rather than hands-on experience of nature. In fact, the number of ways in which theology could be of relevance to the scientific enterprise in the seventeenth century and later is surprisingly large. A belief, for example, in the ultimate unity of nature was often grounded in a monotheistic account of creation, as with Isaac Newton whose convictions about the omnipresence of a single deity informed his belief that his law of gravitation would prove to be of universal application.

To study the changing relations between scientific and religious beliefs has additional value because it can help to correct popular perceptions of the innovations. Two examples will have to suffice, but they relate to two of the major transformations in the history of science: the changes wrought by Copernicus and by Darwin. With respect to the former, how often it is said that the human race was effectively deposed from its central place in creation—diminished in stature as the earth was relegated to a planet among others. For Freud this was the first of the three great revolutions that had cost humankind its exalted place in nature, the other two being the Darwinian (as humans were rendered continuous with the animal creation) and his own in which we became no longer masters of our own minds. But was it really like that?

In Aristotelian philosophy it was the earth that was corrupt and subject to change; the heavens, the regions above the moon, were immaculate and immutable. Objects there had the perfection of cyclic motion that was unrealisable on earth. For Copernicus to place the earth among the planets, far from a denigration, was to elevate it into a superior region. Galileo could speak of our escape from the refuse bin where everything fell back to earth. For Kepler there was exhilaration and elevation in our privileged place in the cosmos, occupying the central orbit in the planetary system. The English Copernican John Wilkins once said that the objection he often heard to the new astronomy was that it elevated humanity above its true station—not that we were demeaned by the transformation. There were other kinds of elevation, too. In surpassing the knowledge of the ancients, there was a new triumph for the human intellect. Moreover, an aesthetic appreciation of the greater elegance of the Copernican universe over the Ptolemaic had the effect of elevating aesthetic sensibility in the human mind. Some of these considerations became less cogent as the Aristotelian cosmos finally collapsed, but it is a serious mistake to project our secular predispositions back onto an earlier period.

The Darwinian 'revolution' is an even more complex case because of the enormous diversity of the religious responses, ranging from outright rejection to immediate acceptance. Standard views correctly point out that serious issues were raised for Christian theology, some of which have still not been adequately resolved. But it would be a mistake to imagine that churchmen were uniformly hostile. Some saw distinct theological advantages in a unified process of evolution. Might not Darwinism assist with the problem of theodicy: how to rationalise the presence of so much pain and suffering in nature? If competition and struggle were the driving forces of natural selection, then without them there would have been no evolutionary development culminating in human intelligence and responsiveness. One could see the pain as a concomitant of what Asa Gray called one great economical process. Even the most gruesome of creatures might be seen not as a direct creation of a seemingly culpable deity but as an indirect product of a process in which it had been possible for evolution to achieve higher forms, the 'greatest good', in Darwin's view, that we can conceive. Darwin was often perceived as a foe and, in some quarters, still is. But there were theologians of the nineteenth century such as Aubrey Moore in Oxford who claimed that, under the guise of a foe, Darwin had done the work of a friend. He had helped to emancipate Christianity from infantile images of a magician god who had created each species by special intervention. Such a semi-deistic position

The Changing Relations between Science and Religion

was, for Moore, a travesty of a true theological understanding in which the deity was active in everything. The trouble with the God of separate creation was that 'He' could so easily be deemed inactive except when intervening. Moore saw in an evolutionary process the immanent activity of a Creator still at work in the world.

I am not suggesting that any of the above responses to change should be construed as definitive; but they do reveal the diversity of cultural meaning that can be extracted from or imposed on a scientific innovation. This is a crucial lesson because so often new forms of science are trumpeted in ways that imply they had a single set of implications—that they triumphed in some way over religious or superstitious beliefs. But to suggest that a particular scientific theory *entails* one interpretation of its cultural meaning over all others is a blinkered and historically ill-informed enterprise. Even the wisest among the popularisers sometimes fall into the trap. In his sensitive book *God's Funeral*, AN Wilson, speaking of Darwinism, refers to atheism as an 'unavoidable consequence' of Darwin's theory. But how unavoidable when, by Wilson's own admission, this was not even Darwin's own position?

We should be constantly on our guard against claims that particular theories imply or entail materialism, or the validity of theism for that matter.

There is a radical approach to the whole question of the 'relations between science and religion', which is to say that the two should have nothing to do with each other. If this view were adopted it would mean that there might be a meta-level position from which one could say that the so-called 'relations' have never changed because, strictly speaking, there have never been any worthy of attention. Such a radical separationist position is by no means new, but it has recently been reaffirmed by Stephen Jay Gould in *Rocks of Ages*.

An attraction of the position is that it may relieve tension between communities fighting over the same intellectual territory. But for the historian it is not a satisfying refuge. One has to ask about the pressures that lead to such a stance. It must not be deprived of context. One that has recurred is when scientific thinkers who may also belong to an identifiable religious tradition find themselves under attack from their respective churches for their supposedly threatening science. It was in such a context that Galileo saw value in separation, citing the aphorism that the Bible teaches us how to go to heaven not how the heavens go. When the nineteenth century Catholic evolutionist St George Mivart began to lose the approbation of his church (over doctrinal issues and not principally over his evolutionary science) he pleaded for a separation and saw himself as a latter-day Galileo.

But there are converse contexts in which a thesis of separation has served the interests of a theology seeking to protect itself from further scientific encroachment. An example, again from the nineteenth century, would be the Oxford Professor of Geometry, Baden Powell, who in his contribution to *Essays and Reviews* (1860) not only drew attention to the revolution Darwin was about to effect, but insisted that theology's best recourse was to restrict itself to the moral sphere. Then there need be no further territorial squabbles of the kind that had made progress in introducing the sciences into Oxford so difficult and which had led to progressive retrenchment as the historical sciences had claimed so much holy ground.

We may surely ask what have been the pressures that have induced Gould to seek such a sharp delineation between the two magisteria, assigning the world of 'facts' to the sciences and that of moral sensibility to religion. One might have been the need to reassure the creationists (who have both attacked and used him) that they have nothing ultimately to fear from Darwinian evolution. Another might be the need to show up the deficiencies of those who would banish religions altogether in the name of science.

Having introduced the importance of context, I shall conclude by noting how changes of context can play tricks of

The Changing Relations between Science and Religion

which we need to be aware. My example concerns the use made not of Darwin's theory but that of his French predecessor Lamarck. I owe the example to Peter Bowler who has observed that in England during the 1830s—a decade of reform and political agitation—Lamarck's theory of the transformation of species was both known and seen as subversive of established religion and social values. This was for obvious reasons: humankind might still be at the top of an evolutionary tree, but human dignity was seemingly compromised by our emergence from lower forms of life. Lamarck's transformism was vigorously attacked even by so innovative a geologist as Charles Lyell in his *Principles of Geology* (1830–33). Projected into the world of socio-political discourse, Lamarck's principle of the inheritance of acquired characteristics could also be seen as promoting purely humanistic models of social progress. But later in the century, in the post-Darwinian controversies, Lamarckism returned as a friend to supporters of theistic evolution, because his mechanism for evolution caused fewer problems for the defence of an overarching teleology than Darwin's conception of natural selection. In a changed context, a once outlawed theory became a welcome resource for the defence of a providential God. The Lamarckian internal drive towards complexification could be reinterpreted as of divine origin, leading to the convergence of organic forms as the ascending escalator of life reached its summit. Until well into the twentieth century, as Bowler has recently shown, schemes of theistic evolution were commonly more Lamarckian than Darwinian in inspiration.

Such stories remind us that we, too, are creatures of context and that we have no way of escaping the historical flux. To ask why we believe what we do about the mutual bearings of our scientific and religious commitments (or their absence) is to ask a question each of us could only answer thoroughly in an autobiographical, historical mode. Whether an astute historian would give the same account as we ourselves is a nice question.

John Hedley Brooke

Bibliography

Bowler, Peter J, *The Eclipse of Darwinism: Anti-Darwinian Evolution Theories in the Decades Around 1900* (Baltimore: Johns Hopkins University Press, 1983).

Bowler, Peter J, *Reconciling Science and Religion: The Debate in Early Twentieth-Century Britain* (Chicago: University of Chicago Press, 2001).

Brooke, John H, *Science and Religion: Some Historical Perspectives* (Cambridge: Cambridge University Press, 1991).

Brooke, John H, '"Wise Men Nowadays Think Otherwise": John Ray, Natural Theology and the Meanings of Anthropocentrism', *Notes and Records of the Royal Society*, 54 (2000): 199-213.

Brooke, John H, 'Science and Secularisation', in Linda Woodhead, editor, *Reinventing Christianity* (Aldershot: Ashgate, 2001), 229-338.

Brooke, John H and Cantor, Geoffrey N, *Reconstructing Nature: The Engagement of Science and Religion* (Edinburgh: T&T Clark, 1998).

Corsi, Pietro, *Science and Religion: Baden Powell and the Anglican Debate, 1800-1860* (Cambridge: Cambridge University Press, 1988).

Danielson, Dennis R, 'The Great Copernican Cliché', *American Journal of Physics* 69 (2001): 1029-35.

Desmond, Adrian J, *The Politics of Evolution* (Chicago: University of Chicago Press, 1989).

The Changing Relations between Science and Religion

Desmond, Adrian J and Moore, James R, *Darwin* (London: Michael Joseph, 1991).

Fantoli, Annibale, *Galileo: For Copernicanism and For the Church*, translated by George V Coyne (Vatican: Vatican Observatory, 1994).

Force, James E, *William Whiston: Honest Newtonian* (Cambridge: Cambridge University Press, 1985).

Force, James E, and Popkin, Richard H, editors, *Newton and Religion: Context, Nature and Influence* (Dordrecht: Kluwer, 1999).

Gould, Stephen J, *Rocks of Ages: Science and Religion in the Fullness of Life* (New York: Ballantine, 1999).

Harrison, Peter, *The Bible, Protestantism and the Rise of Natural Science* (Cambridge: Cambridge University Press, 1998).

Lindberg, David C, 'Science and the Early Church', in David C Lindberg and Ronald L Numbers, editors, *God and Nature: Historical Essays on the Encounter Between Christianity and Science* (Berkeley and Los Angeles: University of California Press, 1986), 19-48.

Moore, James R, *The Post-Darwinian Controversies* (Cambridge: Cambridge University Press, 1979).

Nesteruk, Alexei V, 'Patristic Theology and the Natural Sciences', Parts 1 and 2, *Sourozh: A Journal of Orthodox Life and Thought*, no. 84 (2001): 14-35 and no. 85 (2001): 22-38.

Peacocke, Arthur R, 'Biological Evolution and Christian Theology—Yesterday and Today', in John R Durant, editor, *Darwinism and Divinity* (Oxford: Blackwell, 1985), 101-30.

Rudwick, Martin JS, 'The Shape and Meaning of Earth History', in David C Lindberg and Ronald L Numbers, editor, *God and Nature*, 296-321.

Rupke, Nicolaas A, *The Great Chain of History: William Buckland and the English School of Geology, 1814-1849* (Oxford: Oxford University Press, 1983).

Shortland, Michael, editor *Hugh Miller and the Controversies of Victorian Science* (Oxford: Oxford University Press, 1996).

Webster, Charles, *The Great Instauration: Science, Medicine and Reform 1626-60* (London: Duckworth, 1975).

Wilson, AN, *God's Funeral* (London: John Murray, 1999).

'God's Two Books': Revelation, Theology and Natural Science in the Christian West

Peter J Hess

For the invisible things by him from the creation of the world are clearly seen, being understood by the things that are made. – Paul, Romans 1:20

Hence there are two books given to us by God, the one being the book of the whole collection of creatures or the book of nature, and the other being the book of sacred Scripture. – Raimundus Sabundus

1. Introduction

1.1 Historiography and models

Human language, as a medium of communicating our experience in every sphere, is seldom intended to be understood merely literally. Indeed, the use of metaphor is central to the language we use in realms as widely diverse as religion and science. And in interpreting our human history we consciously employ metaphors and models, discovering a past that is coloured by our experience precisely as we would expect it to be. That is to say, we interpret our history through a manifold conditioned by factors such as popular tradition, established historiography, and the cultural assumptions of our own society. In the case of the Judeo-Christian West, this interpretive manifold increasingly has reflected the assumptions of our technologically- and scientifically-oriented society. Modern Western science has since 1850 come to be regarded more and more as normative, and even as enjoying an epistemologically privileged position.

When examining the historical relationship between religion and science in the West, we must bear in mind that for more than a century the model dominating its interpretation has been the 'warfare' model. First crafted in the nineteenth century in the works of John William Draper and Andrew Dickson White, the popular misconception that science and religion are natural enemies is regularly reinforced by the repetition of historical portrayals of these disciplines as having been at each other's throats since time immemorial.[1] Fortunately, considerable work has been done in the last two decades to redress the balance and construct a more sophisticated understanding of the complex relationship between religion and science.[2]

---

1. The works in question are, first, James Draper's, *History of the Conflict between Religion and Science* (1874), a polemical book denouncing Christian theologians for their campaign of obscurantism against scientific truth dating back to the church of the first century. Second, Andrew Dickson White's *History of the Warfare of Science with Theology in Christendom* (1870-96) paints a dramatic tableau of what he understood to have been a perpetual state of conflict between science and religion, placing Christianity in the villainous role of relentless saboteur of scientific progress.

2. Prominent among them are David Lindberg and Ronald Numbers's *God and Nature* (Berkeley, 1984) and its forthcoming successor *Science and the Christian Tradition: Twelve Case Histories*, David C Lindberg and Ronald L Numbers, editors, (Chicago: University of Chicago Press, 2001), John Brooke's *Science and Religion: Some Historical Perspectives* (Cambridge, 1991), John Brooke and Geoffrey Cantor's *Reconstructing Nature* (Edinburgh, 1998), Gary Ferngren's *Encyclopedia of Science and Religion in the Western Tradition* (Garland, 2000), and Maggie Osler's *Rethinking the Scientific Revolution* (Cambridge, 2000). Of course, the opposite extreme of constructing a counter-mythology of general harmony and cooperation between science and religion must also be scrupulously avoided.

'God's Two Books'

1.2 The knowledge of God in Western traditions

That science and religion have enjoyed a more complex association than one merely of conflict is made clear by the history of the metaphor most central to the natural theological tradition, that of 'God's two books'. 'Natural theology' is a branch of theology in the Western tradition that historically has been regarded by many as complementary to revealed doctrine. What we know about God is derivable from two coordinated sources: the created world and the revealed Scriptures, or 'the book of nature and the book of Scripture'.[3] The idea was used by many early modern English natural theologians as shorthand for the assumed validity of the design argument for the existence of God. Thomas Browne notes in *Religio Medici*:

> Thus are there two books from whence I collect my divinity: besides that written one of God, another of his servant, nature, that universal and public manuscript that lies expansed unto the eyes of all. Those that never saw him in the one have discovered him in the other. This was the Scripture and theology of the heathens: the natural motion of the sun made them more

---

3. There have been significant treatments of particular sections of this theme. Ernst Robert Curtius, in *European Literature and the Latin Middle Ages*, translated by Willard R Trask (New York: Pantheon Books, 1953), treats the development of the idea in Middle Ages to Renaissance. Among the best discussions of the early modern context—to which I am greatly indebted—is *The Bible, Protestantism, and the Rise of Modern Science*. (Cambridge, 1998). Other treatments are James J Bono, *The Word of God and the Languages of Man: Interpreting Nature in Early Modern Science and Medicine* (Madison: University of Wisconsin Press, 1995), and Frank E Manuel, *The Religion of Isaac Newton* (Oxford: Clarendon Press, 1974).

admire him than its supernatural station did the children of Israel.[4]

Although Browne emphasises his care not to lose sight of the distinction between nature and God, he notes that Christians have tended to 'cast a more careless eye on these common hieroglyphics, and disdain to suck divinity from the flowers of nature'.

To a historian of ideas a number of intriguing questions are raised by these references to the 'two books'. First is the matter of the origins of the metaphor, which seventeenth century English authors ascribe variously to Chrysostom, Ambrose, Augustine, or Lactantius (unfortunately usually without benefit of citing the source). But in addition to this intriguing problem of the origins and transmission to English thought of the theme of God's two books of nature and Scripture, a number of questions about the theme remain unclear. What was the history of its use and development in ancient, medieval and early modern theology and natural science? When did the idea finally fall into neglect, and in response to what circumstances? This article will endeavour to answer these questions by reconstructing the sources and development of the commonplace theme of 'God's two books of nature and Scripture' in Western intellectual history.[5]

---

4. Thomas Browne, *Religio Medici,* edited by James Winney (Cambridge, 1983), part I, section 16 (18-19).
5. The context out of which this paper has grown is a larger project on the responses of seventeenth and eighteenth century natural theologians to the scientific discoveries proceeding out of a protracted scientific revolution. See Peter MJ Hess, *Nature and the Existence of God from Hooker to Paley 1597-1802* (PhD dissertation, Graduate Theological Union, Berkeley, 1993). This work examines the myriad ways in which theologians appropriated, misappropriated, rejected, or otherwise responded to such new ideas as heliocentrism, the theory of atomism, probability and certainty, and the religious implications of microscopic life.

'God's Two Books'

2. Birth of the 'two books' metaphor

Human spiritual and religious life has historically been predicated upon the premise that God can be known through divine revelation. In primal religious traditions, stories about sacred times and sacral space, about the ancestors, and about manifestations of the divine through weather, the seasons, and personal intervention, are transmitted orally across countless generations. Revelation comes through the all-encompassing reality of the natural world, in which people are consciously embedded as part of a unified, living, spiritual ecology.

The development of written communication brought important modifications to the notion of revelation. Oral tradition as the primary mode of transmission across the generations became replaced by the mode of fixing truth in written texts. Religious wisdom could now be recorded in permanent form, and the immediacy of myth passed down by word of mouth was replaced by the mediating text, just as the verbal transmission of moral norms became replaced by their codification into law. With the advent of the Abrahamic faiths, the idea of revelation took on a significantly new character. Although the Hebrew Scriptures, the New Testament, and the Qur'an are in important respects substantially different literary genres from each other, they are all nevertheless understood by their followers to be transmitting the very word of God. The 'book' became of paramount importance in these religions, and it is natural, therefore, that it is in the Judeo-Christian tradition that we can trace the origins and development of the 'two books' metaphor.

2.1 Scripture

There are clear scriptural precedents for the idea that God may be known through both divine word and divine works. There are hints of the metaphor in the Hebrew Scriptures, both in the Psalms and in the later books. Psalm 19:1 majestically articulates a theme that would remain common currency throughout the history of natural theology: 'The heavens declare the glory of God, and the firmament proclaims his handiwork'. The Book of Wisdom, composed in the middle of the first century BCE, articulates the idea that God is known through the divine works, even by Gentiles who have not enjoyed the benefit of revelation (Wisdom 11:6-9). The argument is essentially a privative one, removing excuse from non-believers, who may, by analogy with a human author, contemplate the author of creation through the grandeur and beauty of creatures. This argument initiated a strain of thought persisting at least until the time of Calvin, in whose theology humanity is regarded as standing without excuse before the awesome justice of God.

The New Testament *locus classicus* for the natural knowledge of God, the Pauline declaration in Romans, builds upon this Hebrew idea of the authorship of nature:

> For what can be known about God is plain to them, because God has shown it to them. Ever since the creation of the world his invisible nature, namely, his eternal power and deity, has been clearly perceived in the things that have been made. (Romans 1:19-20)

But if its roots of this theme are Hebrew, the particular intensity of its Pauline expression seems to reflect the confrontation of first century Jewish religious thought with the cultural life of the Roman Empire, and particularly with Hellenistic philosophical categories.

2.2 Patristic thought

There are clear precedents for important elements of the 'two books' metaphor scattered throughout Patristic literature. Justin Martyr's second-century apologetic in his *Second Apology* builds on the *logos spermatikos* notion of Stoicism, the idea that the world is permeated by seeds of the divine Word (Chapter 8).

'God's Two Books'

And Irenaeus of Lyons (c 130–202) in *Adversus haereses* (AD 180/199) provides two essential ingredients of the theme: the works and the word of God: 'One God formed all things in the world, by means of the Word and the Holy Spirit; and although he is to us in this life invisible and incomprehensible, nevertheless he is not unknown; inasmuch as his works do declare him, and his Word has shown that in many modes he may be seen and known'. (Book IV, Chapter 20). The first significant thinker of the Latin Church, Tertullian (c 160–c 225), prefigures the metaphor by suggesting that 'God has from the beginning of all things given as primary witnesses for the knowledge of himself, nature in her [manifold] works, kindly providences, plagues, and indications of his divinity'. Tertullian regards these evidences as counterparts to Scripture, and claims that because his heretical opponent Marcion has rejected most of Scripture, he cannot provide a counterpart in revelation to the knowledge of God we derive from nature.[6] Origen (c 185–254), suggests in the *Philocalia* that 'he who believes the Scripture to have proceeded from him who is the author of nature, may well expect to find the same sort of difficulties in it as are found in the constitution of nature'.

But the clearest instance we have to a formal Patristic statement of the metaphor of 'two books' may be found in St John Chrysostom's (c 347–407) *Homilies to the People of Antioch*. An eloquent Antiochene preacher known as 'John the Golden Tongued', Chrysostom implies in Homily IX that 'nature' serves the function of a book of revelation:

> If God had given instruction by means of books, and of letters, he who knew letters would have learnt what was written, but the illiterate man would have gone away without receiving any benefit . . . This however cannot be said with

---

6. Tertullian, *Adversus Marcionem*, edited and translated by Ernest Evans (Oxford: Clarendon Press, 1972), Book V 5, 539.

> respect to the heavens, but the Scythian, and Barbarian, and Indian, and Egyptian, and every man that walks upon the earth, shall hear this voice; for not by means of the ears, but through the sight, it reaches our understanding . . . Upon this volume the unlearned, as well as the wise man, shall be able to look, and wherever any one may chance to come, there looking upwards towards the heavens, he will receive a sufficient lesson from the view of them . . . [7]

This passage by Chrysostom aptly summarises what we might term the Patristic attitude towards the complementarity of natural and revealed theology. There is ample evidence that the ingredients of the 'two books' metaphor were available in the Patristic period. However, it would not be until the high Middle Ages that the metaphor would reach full articulation with the progressive rediscovery of Aristotelian natural philosophy. The 'two books' would become the primary model for expressing a mature binary epistemology of revelation.

3. Establishment and expansion of the 'two books'
3.1 Medieval efflorescence of the trope
The metaphor of God's two books became firmly established in the high Middle Ages. Alain of Lille's judgment was influential: *'Omnis mundi creatura/Quasi liber et pictura/Nobis est et speculum'*

---

7. Chrysostom, Homily IX 5, 162-63. *The Homilies of S. John Chrysostom, on The Statutes, to the People of Antioch. A Library of the Father of the Holy Catholic Church* (Oxford: Parker, 1842). See also Chrysostom *Homily* X.3, 175: 'We were enquiring how, and in what manner, before the giving of the Scriptures, God ordered his dispensation toward us; and we said, that by the creature he instructed our race, stretching out the heavens; and there openly unfolding a vast volume, useful alike to the simple and the wise, to the poor and the rich, to Scythians and to barbarians, and to all in general who dwell upon the earth . . . '

'God's Two Books'

(Every creature is to us like a book and a picture and a mirror.) Hugh of St Victor regarded both the creation and the incarnation as 'books' of God, and compared Christ as primary revelation to a book.[8] Bonaventure (1217–74) noted in his *Collations on the Hexaemeron* that sensible creatures are 'a book with writing front and back', spiritual creatures are 'a scroll written from within', and Scripture is 'a scroll written within and without'.[9] We should note that Bonaventure's writing illustrates the plasticity of metaphors: here we have three books, and while the two books would become the norm in our particular model, a multiplicity of books would crop up repeatedly in the next six hundred years.

The fullest articulation of the metaphor in medieval philosophy and literature may be found in Raimundus Sabundus's *Theologia Naturalis sive Liber Creaturarum* (1436) (*Natural Theology, or Book of the Creatures*). Sabunde was a Barcelona native, a graduate in arts and medicine, who died prematurely at Toulouse while on his way to take up a post at Paris. His statement of the metaphor, which would become the *locus classicus* for early modern appropriators of the theme:

> Hence there are two books given to us by God, the one being the book of the whole collection of creatures or the book of nature, and the other being the book of sacred Scripture. The first book was given to human beings in the beginning, when the universe of creatures was created, since no creature exists that is not a certain letter, written by the finger of God, and from many creatures as from many letters is composed one book, which is called the book of the creatures. Within this book is included humanity itself, and human beings are

---

8. Ernst Curtius, *European Literature and the Latin Middle Ages*, 219, 320.
9. Bonaventure, *Collations on the Hexaemeron*, 12.14-17

> the first letters of this book. But the second book, Scripture, was given to human beings secondarily to correct the deficiencies of the first book, which humanity could not read because it is blind. The first book is common to everyone, but the second book is not common to all, because only clerics are able to read what is written in it.[10]

A number of interesting points are suggested by this passage. For one thing, the idea that we ourselves are letters in the book of nature carries the intriguing implication that the book of nature can never in its entirety be deciphered, since it will never cease being written, at least not until the eschaton. For another, the idea that the book of Scripture is limited to the interpretation of clerics would ultimately be antithetical to the Protestant confidence that the book of Scripture is open to reading and interpretation by all.

But perhaps the most serious consequence—and what led to the condemnation of Sabunde's work as heretical—was his incautious insistence that the book of Scripture is less accurate. *Theologia Naturalis* was placed on the Roman Index of Prohibited Books in 1595, a victim both of the frigid ideological climate of the Counter Reformation, and of Sabunde's own incautious exaltation of the 'Book of Nature'. As Clement Webb noted in his *Studies in the History of Natural Theology*, the problem with Raymond of Sabunde's articulation of the two books metaphor was not its glorification of the Bible at the expense of tradition, but rather an assertion of the pre-eminent importance of natural knowledge, in the spirit of Roger Bacon.[11] Sabunde's confidence in the 'book of nature' was

---

10. Raimudus Sabundus, *Theologia Naturalis Seu Liber Creaturarum*. Reprint of the Sulzbach edition of 1852, with critical notes, (Stuttgart-Bad: Cannstatt, 1966), 35-36.
11. Clement CJ Webb, *Studies in the History of Natural Theology* (Oxford: The Clarendon Press, 1915), 295-96.

supreme: this science is accessible alike to laymen and to clerks and to every condition of men and can be had in less than a month and without trouble, nor to possess it need one have learned anything by heart or keep any written book . . . and so in the order of our procedure it comes before Holy Scripture (296).

3.2 Humanism, printing, and the Reformation
There were influences at work in both the Renaissance and Reformation movements that would exercise profound impacts on the idea of the complementary books of nature and Scripture. Renaissance scholars initiated the subjection of literary texts—ecclesiastical as well as secular—to intense critical scrutiny, and it would not be many centuries before the application of their methods would extend even to biblical texts. Such textual criticism would inevitably undermine both Scripture and the received texts of classic natural philosophy as authoritative books. The technological innovations leading to the Western invention of mechanical printing in 1543 also created a sea change in the literary world. Paradoxically, printing would also serve to weaken the hold which 'science' texts had held on the medieval mind. Theophrastus Phillippus Aureolus Bombastus von Hohenheim (1493-1541)—Paracelsus for short—is a good example. Peter Harrison suggests that

> He [Paracelsus] took a lead in refashioning the medieval metaphor, contrasting the book of nature with both the Scriptures and the writings of ancient authorities. In place of Galen, Avicenna, and Aristotle, Paracelsus set Nature—that library of books which 'God himself wrote, made, and bound'. Every country, he insisted, is a page of nature's book, and 'he who would explore her must tread her books with his feet'. Scripture is

explored through its letters, but nature from land to land.[12]

The implications of this empirical approach to nature—however much it may have been a literary conceit—would be far-reaching for science, and also for the idea of nature as a book.

The emphasis placed by the Protestant Reformers on Scripture would likewise have a significant impact on the development of the theme. For Luther a sober appreciation of the primary meaning of language as literal rather than allegorical, and as the medium by which God's word is accessible to all, would cut through the wild profusion of 'meanings' and 'signatures' that medieval people had found in the book of nature. John Calvin shared this emphasis on the literal sense of Scripture with Luther, and was if anything even more acutely conscious than Luther of the fallen nature of humanity. Appealing to the Pauline notion that we are without excuse for the knowledge of God, he writes in the *Institutes of the Christian Religion*:

> Therefore, though the effulgence which is presented to every eye, both in the heavens and on the earth, leaves the ingratitude of man without excuse, since God, in order to bring the whole human race under the same condemnation, holds forth to all, without exception, a mirror of his Deity in his works, another and better help must be given to guide us properly to God as Creator.

The two books are not equal partners, in Calvin's theology; rather, the revealed word of God in Scripture is a necessary

---

12. Harrison, *The Bible, Protestantism, and the Rise of Modern Science*, 194-95.

'God's Two Books'

corrective to the deficiencies of nature.[13] The Reformed tradition retained this Calvinist interpretation of the two books, and in the *Belgic Confession* adopted by the Dutch Reformed Church, we read in article 2, 'The Means by Which We Know God':

> We know him by two means: First, by the creation, preservation, and government of the universe, since that universe is before our eyes like a beautiful book in which all creatures, great and small, are as letters to make us ponder the invisible things of God: his eternal power and his divinity, as the apostle Paul says in Romans 1:20. All these things are enough to convict men and to leave them without excuse. Second, he makes himself known to us more openly by his holy and divine Word, as much as we need in this life, for his glory and for the salvation of his own.

3.3 The scientific revolution

The centuries that saw the development of the anti-Aristotelian 'new philosophy' are among the most complicated in terms of tracing the ramifications undergone by the 'two books' metaphor. Descartes (1596–1650) appears to have rejected the

---

13. John Calvin, *Institutes of the Christian Religion*, Book I, Chapter 6, Part 1. 'Not in vain, therefore, has he added the light of his Word in order that he might make himself known unto salvation, and bestowed the privilege on those whom he was pleased to bring into nearer and more familiar relation to himself . . . For as the aged, or those whose sight is defective, when any books however fair, are set before them, though they perceive that there is something written are scarcely able to make out two consecutive words, but, when aided by glasses, begin to read distinctly, so Scripture, gathering together the impressions of Deity, which, till then, lay confused in our minds, dissipates the darkness, and shows us the true God clearly'.

book of nature, since it necessitated an empiricism that was inimical to his rationalist epistemological project. Pierre Gassendi (1592–1655), on the other hand, saw purpose in all of nature, and suggested to Descartes that if he wanted to prove the existence of God, he ought to abandon reason and look around him, that the two books were not to be kept on separate shelves.[14] Johannes Kepler considered astronomers to be priests of God in the book of nature, not surprising if Kepler 'elevated nature as a revelation of God to a status equal to that of the Bible'.[15]

Although for Francis Bacon (1561–1626) the two books seem in practice ultimately to have been kept on separate shelves, in *The Advancement of Learning* he articulates their essential connection:

> Our Saviour lays before us two volumes to study, if we will be secured from error: first, the Scriptures, revealing the Will of God; and then the creatures expressing his power; whereof the latter is a key unto the former: not only opening our understanding to conceive the true sense of the Scriptures, by the general notions of reason and rules of speech; but chiefly opening our belief, in drawing us into a due meditation of the omnipotency of God, which is chiefly signed and engraven upon his works. This much therefore for

---

14. Willim Ashworth, 'Catholicism and Early Modern Science', in *God and Nature*, 140.
15. Richard Westfall, 'The Rise of Science and the Decline of Orthodox Christianity', in *God and Nature*, 220. See Kepler's letter to Fabricius, 4 July 1603, in *Werke* 14:421.

'God's Two Books'

divine testimony and evidence concerning the true dignity and value of learning.[16]

James Bono suggests that Bacon, in a sense, distinguishes between two aspects of the 'Word of God': 'The first is revelatory of God's intentions for man, for his salvation and redemption . . . The second aspect of the "Word of God" is his creative word, that "word" that is productive of the created order itself—of "God's Works"! Through knowledge of God's works, humanity can "read" the book of nature, uncovering the signatures in nature, and beginning to reconstruct this second aspect of the "Word of God"—the language of nature known to Adam in his earthly paradise'.[17]

The theme plays an important role in the thought of Galileo, particularly in his *Letter to Grand Duchess Christina*. As is well known, Galileo argued that the book of nature is written in the language of mathematics, not only implying that mathematics is the sublimest expression of the world because divine, but de facto restricting its full comprehension to those who are appropriately educated. It is worth quoting at length:

> And to prohibit the whole science [of astronomy] would be but to censure a hundred passages of holy Scripture which teach us that the glory and greatness of Almighty God are marvelously discerned in all his works and divinely read in the open book of heaven. For let no one believe that reading the lofty concepts written in that book leads to nothing further than the mere seeing of

---

16. Bacon, *The Advancement of Learning*, quoted in Marshall McLuhan, *The Gutenberg Galaxy: the Making of Typographic Man* (Toronto), 187.
17. James Bono, *The Word of God and the Languages of Man: Interpreting Nature in Early Modern Science and Medicine* (Madison: University of Wisconsin, 1995), 219.

> the splendor of the sun and the stars and their rising and setting, which is as far as the eyes of brutes and of the vulgar can penetrate. Within its pages are couched mysteries so profound and concepts so sublime that the vigils, labors, and studies of hundreds upon hundreds of the most acute minds have still not pierced them, even after continual investigations for thousands of years.

Galileo elsewhere suggested that Scripture teaches us 'how the heavens go and not how to go to heaven', seemingly endorsing a two-languages model—or in Barbourian terms, the independence model—of the science/religion relationship.

The 'two books' metaphor flourished in the natural theological climate of seventeenth-century England, but its two terms were not always held in comfortable balance. The dissenting theologian Richard Baxter, for example, felt that 'nature was a "Hard Book" which few could understand. It was therefore safer to rely more heavily on Scripture'.[18] Newton's thought on the relationship between science and religion is notoriously complex, and interpreters differ considerably about it. Richard Westfall suggests that for Newton nature was perhaps more truly the source of divine revelation than was the Bible, noting that Newton adds 'revelation' almost as an afterthought.[19] On the other hand, Frank Manuel argues that Newton—in virtually abolishing the distinction between the two books, which he revered as separate expressions of the

---

18. Richard Baxter, *The Reasons of the Christian Religion* (London: 1667), 193. Quoted in Barbara Shapiro, *Probability and Certainty in Seventeenth-Century England*, 94, quoting John Ray. Shapiro also suggests that 'the emphasis placed upon the two books varied with different authors' (291, note 65).
19. Westfall, 'The Rise of Science', 232-33. Westfall's source is a manuscript in the Jewish National and University Library, Jerusalem, Yahuda MS 41, fols. 6 & 7.

same divine meaning—was attempting to keep science sacred and to reveal scientific rationality in what was once a purely sacral realm, namely, biblical prophecy. Manuel suggests that even Newton was uneasy about the amalgam, and that he was aware that science and its uses were becoming independent of theology 'despite the proliferation of books of *physica sacra* and the depth and pervasiveness of his own religious feelings'[20]

4. Waning and survival of the metaphor
4.1 Challenges: deism and historicism

The theme of the 'two books' persisted vigorously right through the nineteenth century, and there are a number of intriguing (if repetitive) book-length treatments of it. However, we can also begin at this time (around 1800) to see cracks in the edifice of the metaphor. The deist movement challenged the uniqueness of the Christian revelation. One telling example is Thomas Paine, who writes defiantly in his *Age of Reason*:

> But some, perhaps, will say: are we to have no word of God—no revelation? I answer, Yes; there is a word of God; there is a revelation. The Word of God is the creation we behold, and it is in this word, which no human invention can counterfeit or alter, that God speaketh universally to man . . . In sum [fine], do we want to know what God is? Search not the book called the Scripture, which any human hand might make, but the Scripture, called the creation.[21]

---

20. Frank E Manuel, *The Religion of Isaac Newton* (Oxford: Clarendon Press, 1974), 49.
21. Thomas Paine, *Age of Reason, Being an Investigation of True and Fabulous Theology* (Luxembourg, 1794), 38, 41. Other early deist works are John Toland's *Christianity Not Mysterious,* and Mathew Tindal's *Christianity As Old As the Creation.*

Paine admits the possibility of revelation, but rejects the idea that God has ever communicated with humankind otherwise than through the universal display of Godself in the works of the creation, and through the moral sense of repugnance to bad actions and attraction to good ones. Creation is the Bible of the deist:

> Instead then, of studying theology, as is now done, out of the Bible and the Testament, the meanings of which books are always controverted and the authenticity of which is disproved, it is necessary that we refer to the bible of the creation. The principles we discover there are eternal and of divine origin; they are the foundation of all the science that exists in the world, and must be the foundation of theology. We can know God only through his works.[22]

The deist challenge would have a profound impact in philosophical circles, attacking one of the pillars of the 'two books' theme. But there were other trends in the nineteenth century that would exercise an even more widespread effect, including the revolutions in geology and biology which challenged longstanding traditions of a young earth and an immutable creation, and therefore of a coherent 'book of nature' temporally coextensive with the 'book of Scripture'. Charles Babbage (1791-1871) advanced a view in his *Ninth Bridgewater Treatise* (1838) that seems to have verged almost on asserting the superfluity of scriptural revelation:

> In the early stages of the world, before man had acquired knowledge to read the book of nature ever open to his view, direct revelation might be as necessary for his belief in a deity, as for his moral

---

22. *Ibid*, 240, 247, 250-51.

'God's Two Books'

> government; and this might from time to time be repeated. When civilization and science had fixed their abode amongst mankind, and when observations and reason had enabled man to penetrate some little way into the mysteries of nature, his conviction of the existence of a first great cause would gradually acquire additional strength from the use of his own faculties, and when accumulating proofs had firmly established this great step, the recurrence of revelation might be less necessary for his welfare.[23]

Babbage continues by arguing that even if 'the ancient revelation' lost its ability to convince through its transmission through the centuries, modern science can give it a degree of force that can compel 'our understandings to assent to it even with a conviction great as that which had compelled the belief of those to whom it was originally delivered'.

In addition to the 'historicisation' of geology and biology, the nineteenth century saw the development of an historical-critical approach to study of the Bible. This would affect the 'two books' theme no less importantly, challenging the entire received tradition about the nature of Scripture as a unitary record of the word of God. Naturally, these innovations in both hermeneutics and science would push some people in the more conservative wings of society in the pre-critical direction of maintaining verbal inerrancy and defending the ancient understanding of earth history. The metaphor of 'God's two books' of revelation would gain weight as one of the cornerstones of their position.

---

23. Charles Babbage, *The Ninth Bridgewater Treatise, a Fragment* (London: John Murray, 1838; repeat edition (London: Frank Cass & Co, Ltd, 1967), 138-39.

Peter J Hess

4.2 Survivals of the metaphor

The theme of 'two books' would continue to thrive under both conservative and liberal interpretations. Commenting on Genesis 2:15, Seventh Day Adventist leader Ellen G White (1827-1915) expanded upon the theme that Adam and Eve were committed to the care of the garden, 'to dress it and to keep it'. White argued that 'the book of nature, which spread its living lessons before them, afforded an exhaustless source of instruction and delight. On every leaf of the forest and stone of the mountains, in every shining star, in earth and sea and sky, God's name was written . . . all of these were objects of study by the pupils of earth's first school'. The laws and operations of nature, and the great principles of truth that govern the spiritual universe—these were opened to Adam's and Eve's minds by the infinite Author of all.[24]

The theme survived as well among numerous nineteenth-century 'liberal' thinkers, who had by and large adopted the principles and findings of contemporary science. Herbert W Morris argues in *Science and the Bible* (published a decade after Darwin's *Origin*) that Scripture and nature represent respectively the verbal and the pictorial representation of divine wisdom. Morris intended to illustrate the 'Inspired Record of Creation' by reference to the 'marvelous developments of modern science,' taking great pains to obtain the latest and most accurate results of science.[25] Paul A Chadbourne lived and breathed the metaphor in his 1867 Lowell Lectures delivered in Boston, *Lectures on Natural Theology, or Nature and the Bible from*

---

24. Ellen G White, *Education*, Chapter 2, 'The Eden School' (1903).
25. Herbert W Morris, *Science and the Bible, or, the Mosaic Creation and Modern Discoveries* (Philadelphia: Ziegler and McCurdy, 1871), 4-5. Championing a long geological history of the pre-Adamite earth, Morris appears to reject evolution, arguing instead that God creates successive 'races' of creatures as catastrophes reshape the surface of the earth. (42)

'God's Two Books'

the Same Author, and starkly articulated the challenge posed to contemporary science by the book of God's word:

> Above all the sources of knowledge, we have a Book, claiming divine origin, claiming to be the written word of the Being we are searching for, revealing His character and answering every question we need to propound respecting Him and our relations to him . . . but we freely acknowledge that the Bible must stand the tests which science can fairly put it to. If, by fair interpretation, it is shown to conflict with the revelations of nature, it can no longer claim authority as the word of God.[26]

But Chadbourne is curiously out of date. A decade after the appearance of the *Origin of Species*, he has set up for meticulous 'scientific' examination a safe, immutable, pre-Darwinian concept of the world. He championed the idea that nature is an unchangeable record, a 'temple inscribed', contending that geology, chemistry, and biology constitute the language in which nature is written, with geology being the most clearly comprehended volume. It is no wonder that he could confidently conclude that 'the two revelations are one in their teaching'.[27]

In stark contrast is the very progressive writer Joseph Le Conte, professor of geology at the University of California, Berkeley. Writing in an apologetic pro-evolutionary context, LeConte declared that 'the whole object of science is to construct the theology or the divine revelation in nature', and contended that 'while Metaphysics has been disputing as to

---

26. P A Chadbourne, *Lectures on Natural Theology, or Nature and the Bible from the Same Author* (New York: GP Putnam and Son, 1870).
27. *Ibid*, 55, 217-18, 222, 237, 320.

whether watches are made or whether they are eternal, Geology comes forward and tells us the date of manufacture'.[28]

He mounted a vigorous defence of the theory of evolution, but was quite clear on the limits of science as a commentary on the book of nature, which passes from sensible phenomena to immediate causes, and from these to other 'higher causes', and by a continuous chain reaches the Great First Cause, where 'she doffs her robes, lays down her scepter, and veils her face'.[29] For LeConte, 'Of these two books, Nature is the elder born, and, in some sense, at least, may be considered the more comprehensive and perfect'. But he hastens to qualify this: 'Nature cultivate primarily the intellect, while Scripture cultivates primarily the moral nature of man'.[30] The originality of LeConte's careful and detailed exposition of the two books theme is his thorough integration of contemporary science, and his attention to emerging issues in the philosophy of science.

4.3 Eclipse of the metaphor

Despite the healthy survival of the theme of 'God's two books' in evangelical theology and the works of some scientists late in the nineteenth century, during the following hundred years it would fall into a general eclipse. Some prominent examples are instructive in this regard.

Princeton theologian Charles Hodge investigated the issue of the human knowledge of God in a thoroughly systematic way in his *Systematic Theology* (1873), arguing that the mind proceeds to form its idea of God by way of analogy: if we are like God, then God is like us, and the works of God manifest a nature like

---

28. Joseph Le Conte, *Religion and Science: a Series of Sunday Lectures on the Relation of Natural and Revealed Religion, or the Truths Revealed in Nature and Scripture* (New York: D Appleton and Company, 1902), 117, 21.
29. *Ibid*, 23.
30. *Ibid*, 244, 245.

'God's Two Books'

our own.[31] However, he nowhere employed the metaphor, and one wonders whether in the era of a critical approach to Scripture it was becoming considered unsophisticated to appeal to 'God's two books'. Hodge's later Princeton colleague, the influential J Gresham Machen, argued in *The Christian Faith in the Modern World* (1936)—without appealing to the theme—that 'the revelation of God through nature has the stamp of approval put on it by the Bible'.[32]

We find clear evidence for part of the reason for the breakdown of the metaphor in Frederick Temple's *Nature, Man and God* (the Gifford Lectures for 1933-34):

> The supposed clearness of the distinction between Natural and Revealed Religion, as it existed in the minds of our grandfathers, was partly illusory. For us it has, in that form, been completely destroyed by recent study of what has been taken to be the main source of revealed religion—the Bible. In the eighteenth century, and for much of the nineteenth, the theologian believed himself to draw his principles from the lively oracles of God contained in Holy Scripture, and developed his theology as a deductive science. But we are now vividly conscious that whatever the books of the Bible may contain of divine self-disclosure, they

---

31. Charles Hodge, *Systematic Theology* (1873), Volume I repeat edition (London: James Clarke, Ltd, 1960) Chapter IV, 'The Knowledge of God,' 339-45.
32. J Gresham Machen, *The Christian Faith in the Modern World* (New York: Macmillan 1936), 20. More specifically, 'God has been pleased to reveal himself in two ways. In the first place, he has been pleased to reveal himself through nature—by the wonders of the world and by his voice within, the voice of conscience—and, in the second place, he has been pleased to reveal himself in an entirely different way that we call "supernatural" because it is above "nature"'. (32).

are also the record of a very rich and significant human experience.[33]

Temple does not refer in any way to the metaphor of God's two books, suggesting that through the revolutions in scientific method and understanding, and in biblical history and exegesis, the theme of two coordinated revelations of nature and Scripture was well on its way to retirement.

5. Conclusions

The metaphor of 'God's two books' has enjoyed an extraordinarily complex history, from its roots deep in antiquity right up to the present time. Our survey of its intriguing lifecycle leads us to some important conclusions about why the metaphor began to erode and eventually die out in most circles.

First, we have seen that the metaphor was born at the confluence of a number of streams: the common human experience of the transcendent, the conviction of the possibility of divine communication, and the Western fascination for books as repositories of knowledge. As we have seen, the theme underwent a complex process of development, as the natural world was paralleled first by the book of the Scriptures in Hebrew history, then by the increasing number of written records in late antiquity, and later by the printed book in 1543. The production of books on a mass scale with the advent of printing worked a radical change in both the nature of book learning and the character of the book itself. Books became more commonplace and clearly reflected new mechanical techniques of production. Combined with the shift in scholarship from a lifetime of glossing a few volumes to a wide-ranging effort of cross-referencing vast numbers of works, this

---

33. Frederick Temple, *Nature, Man and God* (the Gifford Lectures 1932-34). Lecture I, 'The Distinction between Natural and Revealed Religion', (London: Macmillan, 1935), 5-6.

no doubt contributed to the loss of the quasi-sacred character of the 'book'.

A second factor underlying the atrophication of the metaphor in some circles is what we might term a gradual 'deistic drift'. We have seen that for Thomas Paine the Scriptures testifying to the experience of a particular people's experience of God are no longer trustworthy as the universal book of nature; indeed, perhaps they are not trustworthy at all. Could Scripture defend its place as the Book of all books in a deistic world view?

A third circumstance was the gradual development of a historical-critical approach to biblical interpretation, paralleled by the historical treatment of geology. As the Bible came to be recognised as a compendium of many different genres of writing, assembled over centuries and through numerous and diverse redactions, it became more difficult to read it as a unitary and timeless presentation of the divine word. Scripture came to be understood as a contextualized document proceeding from varied and (often contradictory) human experience. Likewise, we can trace a gradual historicisation of nature through the eighteenth century development of the deep history of time. The geological record showed a progressive revelation, and while this was accommodated reasonably well by some, such as Le Conte, it would not be accommodated well by others.

Parallel to this historicisation was a growing loss of confidence in the idea that we can easily interpret scientific evidence teleologically. Beginning with the eighteenth century discovery of extinction in the fossil record, the evidence mounted that species are not immutable, and that the physical world is not always or even usually the harmonious environment characterised by William Paley's quiet vicarage garden. All this rendered it increasingly difficult to look upon 'the book of nature' as self-evidently revealing the divine plan, or at least as a plan worthy of admiration.

Peter J Hess

A fifth factor behind the decline of the metaphor is the gradual professionalisation of the sciences in the nineteenth century. As natural philosophy split into physics and chemistry and astronomy, and as natural history was divided into biology and ecology and paleontology, each with its own discreet subject matter and unique methodologies, the 'nature' underlying the 'book of nature' lost its metaphorical cogency. Moreover, the virtually complete replacement of commentary on ancient texts by the empirical exploration of the physical world, essentially took the 'book' out of the 'book of nature'. Thus, the metamorphosis of 'natural philosophy' or 'natural history' into the professional disciplines as we know them today undercut both terms in the metaphor of 'God's two books'.

A final factor—particularly relevant in the late twentieth century—might be the changing flow of information consequent upon the computer revolution. We are so inundated with information at all times and from every conceivable source, that the medieval idea of two 'books' as primary sources of revelation seems hopelessly anachronistic. It has even been common to speak since the 1960s of the eventual death of the book as a human artefact of knowledge, replaced by the explosion of the almost infinite and instantaneous transmission of information, now available over the Internet. To be sure, information is hardly the same as revelation or wisdom, but it is a way of obtaining information that appears to out-competing the latter for attention in many spheres.

The complex theme of 'God's two books' has enjoyed a long and convoluted history. Although during the last century it has gradually fallen into disuse, the metaphor has for nearly two millennia variously framed, constituted, negated, or otherwise reflected the relationship between the two human enterprises that would in time become science and religion. In a postmodern era it appears to be a less convincing rhetorical device. For historians of this interaction, however, understanding the lifecycle of the metaphor is pedagogically very useful, since the history of its employment can tell us a lot about how science and religion were viewed.

Can the theme of 'God's two books' be replaced in a world of historical-critical interpretation of Scripture, and in a world in which evolutionary or developmental paradigms hold sway in scientific disciplines ranging from cosmology and geology to biology and neuroscience? Whether or not it can be rehabilitated, the changing fashions of metaphor cannot mask the conviction for believers that God does speak to God's creatures in pluriform ways: through religious traditions, through immediate intuition, through personal relationships, and through the revelations of Scripture and nature.

Bibliography

Primary Sources

Athanasius, *Vita S Antoni* AD 356/362.

Bacon, Francis, *Advancement of Learning* (1605).

Babbage, Charles, *The Ninth Bridgewater Treatise, a Fragment* (London: John Murray, 1838; repeat edition London: Frank Cass & Co, Ltd, 1967).

Balfour, Thomas AG, *Nature and the Bible Have One Author* (London: James Nisbet, 1861).

Baxter, Richard, *The Reasons of the Christian Religion* (London, 1667).

Bonaventure, *Collations on the Hexaemeron* (1274).

Brennan, Martin S, *The Science of the Bible* (St Louis: B Herder, 1898).

Peter J Hess

Browne, Thomas, *Religio Medici* (1635), edited by James Winney (Cambridge, 1983).

Butler, Joseph, *Sermons*.

Calvin, John *Institutes of the Christian Religion*, Book I, Chapter 6, part 1.

Chadbourne, Paul A, *Lectures on Natural Theology, or Nature and the Bible from the Same Author* (New York: G P Putnam and Son, 1870).

Chrysostom, John, *The Homilies of S. John Chrysostom, on The Statutes, to the People of Antioch. A Library of the Father of the Holy Catholic Church*, Homily IX (Oxford: Parker, 1842).

Cicero, Marcus Tullius, *De falsa religione*.

Dante, *Paradiso*, XXIII.

Dick, Thomas, *The Christian Philosopher, or the Connexion of Science and Philosophy with Religion.* (1844).

Dove, John, *A Confutation of Atheism* (London, 1605).

Hodge, Charles, *Systematic Theology* (1873), Volume I. Repeat edition (London: James Clarke, Ltd, 1960).

Irenaeus of Lyons, *Adversus haereses* AD 180/199.

Justin Martyr, *Second Apology*.

Kepler, Johannes, Letter to Fabricius, 4 July 1603, in *Werke* 14:421.

'God's Two Books'

Le Conte, Joseph, *Religion and Science: a Series of Sunday Lectures on the Relation of Natural and Revealed Religion, or the Truths Revealed in Nature and Scripture* (New York: D Appleton and Company, 1902).

Machen, J Gresham, *The Christian Faith in the Modern World* (New York: Macmillan 1936).

Morris, Herbert W, *Science and the Bible, or, the Mosaic Creation and Modern Discoverie* (Philadelphia: Ziegler and McCurdy, 1871).

Newton, Isaac, Yahuda MS 41, fols. 6 & 7.

Paine, Thomas, *Age of Reason, Being an Investigation of True and Fabulous Theology* (Luxembourg, 1794).

Temple, Frederick, *Nature, Man and God* (the Gifford Lectures 1932-34) (London: Macmillan, 1935).

Tertullian, *Adversus Marcionem* edited and translated by Ernest Evans (Oxford: Clarendon Press, 1972).

White, Ellen G, *Education* (Oakland, CA: Pacific Press Publishing Co, 1903).

Secondary Sources

Ashworth, William, 'Catholicism and Early Modern Science', in *God and Nature*.

Bono, James J, *The Word of God and the Languages of Man: Interpreting Nature in Early Modern Science and Medicine* (Madison, University of Wisconsin Press, 1995).

Curtius, Ernst Robert Curtius, in *European Literature and the Latin Middle Ages*, translated by Willard R Trask (New York: Pantheon Books, 1953).

Harrison, Peter *The Bible, Protestantism, and the Rise of Modern Science* (Cambridge, 1998).

Holton, Gerald, 'Johannes Kepler's Universe: Its Physics and Metaphysics', *American Journal of Physics* 24 (May 1956): 340-351.

Lindberg, David C, and Numbers, Ronald L, editors, *God and Nature: Historical Essays on the Encounter between Christianity and Science* (Berkeley: University of California Press, 1984).

McLuhan, Marshall, *The Gutenberg Galaxy: the Making of Typographic Man* (Toronto: University of Toronto Press, 1962).

Manuel, Frank E, *The Religion of Isaac Newton* (Oxford: Clarendon Press, 1974).

Moore, JR, 'Geologists and Interpreters of Genesis in the Nineteenth Century', in Lindberg and Numbers, editors, *God and Nature*.

Osler, Margaret J, editor, *Rethinking the Scientific Revolution* (Cambridge, 2000).

Raven, CE, *Organic Design: A Study of Scientific Thought from Ray to Paley* (Oxford University Press, 1953).

Sabunde, Raymond, *Theologia Naturalis Seu Liber Creaturarum* (1436). Reprint of the Sulzbach edition of 1852, with critical notes, Stuttgart-Bad Cannstatt, 1966.

'God's Two Books'

Shapiro, Barbara, *Probability and Certainty* in *Seventeenth-Century England.*

Webb, Clement CJ, *Studies in the History of Natural Theology* (Oxford: The Clarendon Press, 1915).

Westfall, Richard, 'The Rise of Science and the Decline of Orthodox Christianity', in *God and Nature.*

Part Two

Philosophical Perspectives

Why Christians Should Be Physicalists

Nancey Murphy

1. Introduction

Pick up any reference work from the Christian scholarly world and read what it has to say about the nature of the human person, the body, soul, resurrection, immortality, and it is likely to tell you more about the assumptions of the era in which it was written than it does about original or authentic Christian teaching. There seems to be no topic in theology or biblical studies into which we humans are more likely to project ourselves. This is not surprising. For this reason I shall not presume to tell you about *the* Christian view of human nature.

My plan, instead, is to piece together a brief account of how these issues have developed throughout Christian history. I will end this historical account with current arguments against dualism (or trichotomism) and in favour of a position I will call nonreductive physicalism. Then we need to ask whether such an account should be considered to fall within the range of acceptable positions within the Christian tradition. There are both theological and philosophical issues to consider. I shall touch briefly on some of the theological issues at the end of this paper, but in my second paper I shall concentrate on philosophical issues, attempting to further arguments for the possibility of genuinely *nonreductive* physicalism—that is, to show that neurobiological determinism does not threaten our self-image as free and rational creatures.

2. History

I have failed to discover any comprehensive history of the issue with which I'm concerned here—the metaphysical make-up of the human person. One aspect that needs to be included in any such account is the history of the *oversimplifications* of earlier history—to which I hope I am not now contributing! Here is my amateur historian's account.[1]

Apparently there were a variety of theories of human nature, with correlative expectations regarding death, available to the writers of the New Testament. It is widely agreed among current Christian and Jewish scholars that early Hebraic accounts of the person were holistic and physicalist, and offered no well–developed account of life after death. By Jesus' day, however, there was a lively debate as to whether or not the dead would rise at the end of time. The Hellenisation of the region had begun several centuries earlier and some Jews had adopted a dualistic view of body and soul, along with a conception of the soul's survival of death. Early Gentile Christians probably held an even wider variety of views. The important fact to note is that there is no explicit teaching on the metaphysical composition of the person; however, the New Testament writers did clearly emphasise the resurrection of the body (as opposed to immortality of the

---

1. For a slightly more comprehensive account, see my Introduction in RJ Russell, N Murphy, TC Meyering, and MA Arbib, editors, *Neuroscience and the Person: Scientific Perspectives on Divine Action* (Vatican City State and Berkeley: Vatican Observatory and Center for Theology and the Natural Sciences, 2000), i-xxxv; and Warren S Brown, Nancey Murphy, and H Newton Malony, editors, *Whatever Happened to the Soul? Scientific and Theological Portraits of Human Nature* (Minneapolis: Fortress Press, 1998), especially my chapter 1 and Joel Green's chapter 7.

soul) as the guarantee of life after death. Writing to the church at Corinth, Paul's apology for the resurrection of the body met resistance from some who found it too good to be true and from others who could not understand why they should *want* to be encumbered again by a body once they had escaped it at death.

As Christianity spread throughout the Mediterranean world and its theology was developed in conversation with a variety of philosophical and religious systems, a modified neoplatonic account of the person came to predominate in scholarly circles. The *eternal* Platonic soul became (merely) immortal and there was added the expectation that it would be reunited with a body at the end of time. Augustine's account was the most influential until the later Middle Ages.

A major turning point in Christian history was a result of borrowing from Muslim scholarship in the later Middle Ages. I shall return shortly to Thomas Aquinas's account of the soul, with its dependence on Aristotle and on further developments by Muslim scholars. Thomas's position, based on Aristotle's conception of the soul as the form of the body, may be described as a modified rather than radical dualism.

Two factors at the dawn of modernity challenged the Aristotelian account of human nature. One was the Protestant Reformation's tendency to associate Aristotle with Catholicism and to return to the more Platonic elements in Augustine's thought. The other was the demise of Aristotelian metaphysics as a whole as a result of the rise of modern science—the substitution of atomism for hylomorphism. In response, René Descartes provided modern Europeans with a dualism of mind and body even more radical than Plato's—mental substance is defined *over against* material substance, and the body is purely mechanical.

The interesting twists in this story are the result of critical church history and historical-critical biblical scholarship, beginning especially in the nineteenth century. At that time

Why Christians Should be Physicalists

many scholars called into question the authenticity of miracle accounts in the Bible, and especially the chief miracle, the resurrection of Jesus. This led to an emphasis in theological circles on an immortal soul as the only basis for Christian hope for life after death. Immanuel Kant's transcendental argument for the immortality of the soul played a complementary role.

At the same time, though, critical scholarship made it possible to ask whether current doctrine (including doctrines regarding the soul) were in fact original Christian (and Hebraic) teaching or whether they were the result of later doctrinal development, read back into the biblical texts. It became common during the twentieth century to make a sharp distinction between original Hebraic conceptions and later Greek accretions such as body-soul dualism, and to favor the former as authentic Christian teaching. In addition, both theologians and biblical scholars in the past generation have rediscovered the centrality of the resurrection of the body in primitive Christian proclamation.

Science has affected these debates at three major points. First, as already mentioned, the atomist revolution in physics represented the replacement of Aristotelian hylomorphism—the view that all material things are composed of matter and form. Thus, not only did it become impossible to understand soul as the form of the body, but the very conception of matter involved in speaking of the body changed radically. Second, evolutionary biology pushed many in the direction of physicalist accounts of human nature: if animals have no souls (as moderns, beginning with Descartes, assumed), then humans must not have them either. But others argued that the concept of soul is all the more important in order to account for human distinctiveness. The thesis of this paper is that the most significant scientific development having a bearing on this long history of debates is now occurring in the cognitive neurosciences.

3. Neuroscience and the soul

In this section, then, I shall argue that all the human capacities once attributed to the soul can now be understood as brain functions—more precisely, as functions of the brain in the body and in complex social relations. In order to provide evidence for this claim, however, we need an account of what those soul capacities are. I use Thomas's account for this purpose because, to my knowledge, it is the most elaborate and perceptive analysis in the Christian tradition—he was an excellent cognitive scientist.

Thomas had an elaborate account of the hierarchically ordered faculties of the soul, which derived ultimately from Aristotle's distinction among the capacities of the vegetative, sensitive (or animal), and rational souls. The lowest powers of the human soul, shared with both plants and animals, are the vegetative faculties of nutrition, growth and reproduction. Next higher are the sensitive faculties, shared only with animals, and including the exterior senses of sight, hearing, smell, taste, and touch, and four 'interior senses' (to which I shall return shortly). This sensitive level of the soul also provides for the power of locomotion and for lower aspects of appetite—the ability to be attracted to sensible objects. This appetitive faculty is further subdivided between a simple tendency towards or away from what is sensed as good or evil, and a more complex inclination to meet bodily needs or threats with appropriate responses: attack, avoidance or acquiescence. Together these appetitive faculties (all still at the sensitive level) provide for eleven kinds of emotion: love, desire, delight, hate, aversion, sorrow, fear, daring, hope, despair and anger.

The rational faculties are distinctively human: passive and active intellect and will. The will is a higher appetitive faculty whose object is the good. Since God is ultimate goodness, this faculty is ultimately directed towards God. The two faculties of the intellect enable abstraction, grasping or comprehending the abstracted universals, judging and remembering. Morality is a function of attraction to the good combined with rational judgment as to what the good truly consists in.[2]

The concept of the soul, in general, is an explanatory concept invoked to explain the existence of capacities such as these, since they did not appear to be explainable in physical terms. Now, however, all of them can profitably be studied as biological functions generally or, more particularly, as functions of the brain and nervous system. The soul was, first of all, *anima,* the life force that *anima*ted the body. Since the end of the vitalist controversy early in the twentieth century, though, life has been seen as a function of the organisation of the body, not as the result of an additional, nonmaterial substance. The soul was also taken to account for the fact that members of a species always reproduce their own kinds, with some properties being 'essential' (such as four-leggedness in horses) and others being 'accidental' (such as color). The discovery of DNA, of course, now explains all of this.

The three functions of the vegetative soul, nutrition, growth and reproduction, are all understood now in purely biological terms. The brain is clearly involved in all the higher faculties: neuroscientists have located the motor cortex, auditory and visual cortices, olfactory lobes and so forth. It was once thought that all emotions were mediated by the same neural machinery, the 'limbic system', but more recent research suggests that there are different systems for different emotions.[3]

---

2. Thomas Aquinas, *Summa Theologica,* Ia, 75-83.
3. Joseph LeDoux, *The Emotional Brain: The Mysterious Underpinnings of Emotional Life* (New York: Simon and Schuster, 1998). Leslie Brothers, in fact, argues that the general concept of emotion does not hold up as a useful category in neuroscience. See her *Friday's Footprint: How Society Shapes the Human Mind* (New York: Oxford University Press, 1997), chapter 8.

Nancey Murphy

The functions of the rational soul are less well understood in neurobiological terms than sensations. However, all involve language, and a great deal is known about the brain's role in language use. Broca's area and Wernicke's area have long been recognised as language centers; different regions of the brain have been shown to be involved in recall of different sorts of words—nouns, proper names, verbs.[4] Furthermore, syntax and semantics are processed by different brain systems.[5]

I come back now to Thomas's four 'interior senses', attributed to the sensitive soul. There are precursors of Thomas's views to be found in Aristotle, but largely Thomas borrowed here from Ibn Sina. The *sensus communis* (common sense) is the faculty that distinguishes and collates the data from the exterior senses—for example, associating the sweetness of honey, its color, texture, and scent in order to allow for recognition of the one substance. The *vis aestimativa* (the estimative power or instinctive judgment) allows for apprehensions that go beyond sensory perception, for example, apprehending the fact that something is useful or friendly or unfriendly. The *vis memorativa* (sense memory) stores the judgments made by instinct regarding the intentions of agents. Notice that these philosophers are right: these are complex abilities that we share with animals.

Ibn Sina recognised two further internal senses: the representational power that preserves the sensations of the common sense even after sensible things disappear, and the imagination that selects and combines some of the objects of the representational power with each other and to separate

---

4. See especially Terrence W Deacon, *The Symbolic Species: The Co-evolution of Language and the Brain* (New York: Norton, 1997).
5. See Peter Hagoort, 'The Uniquely Human Capacity for Language Communication', in Russell et al, *Neuroscience and the Person: Scientific Perspectives on Divine Action*, 45-56.

the rest.[6] Thomas did not believe that animals other than humans have the latter ability, and so for him *phantasia* (imagination) is actually Ibn Sina's representational power.[7]

In contemporary neuroscience an explanation for what the medievals called the common sense is now referred to as the binding problem—that is, what accounts for the unity of consciousness—and it is considered one of the most difficult problems in current research, second only to the problem of consciousness itself.

An important question in neuroscience has been the controversy over how the brain comes to recognise patterns. Do brains come equipped with individual neurons designed for recognising patterns—that is, a 'grandmother neuron' devoted to recognition of this one particular elderly woman, and other cells for each pattern that the brain is able to distinguish? It is now believed that recognition tasks depend on activation of large nets or assemblies of neurons rather than on the firing of individual neurons. The concept of a 'cell assembly' was introduced by Donald Hebb, and its formation is described as follows: 'Any frequently repeated, particular stimulation will lead to the slow development of a 'cell-assembly', a diffuse structure comprising cells . . . capable of acting briefly as a closed system . . .'[8] This issue is clearly relevant to an understanding of Thomas's imagination and Ibn Sina's representational power.

It appears that the concept of the estimative power was first developed by Ibn Sina. This is a particularly interesting faculty from the point of view of neuroscientific in-

---

6. Shams C Inati, 'Soul in Islamic Philosophy', in Edward Craig, editor, *Routledge Encyclopedia of Philosophy* (London and New York, 1998), volume 9, 41.
7. Thomas Aquinas, *Summa Theologica,* 1a, 78, 4.
8. Quoted by Alwyn Scott, *Stairway to the Mind: The Controversial New Science of Consciousness* (New York: Springer Verlag, 1995), 81.

vestigations, and it needs to be considered along with the sensitive memory. What Joseph LeDoux writes about 'emotional appraisal' is relevant to distinguishing the estimative power from the common sense:

> When a certain region of the brain is damaged [the temporal lobe], animals or humans lose the capacity to appraise the emotional significance of certain stimuli without any loss in the capacity to perceive the stimuli as objects. The perceptual representation of an object and the evaluation of the significance of an object are separately processed in the brain. [In fact] the emotional meaning of a stimulus can begin to be appraised before the perceptual systems have fully processed the stimulus. It is, indeed, possible for your brain to know that something is good or bad before it knows exactly what it is.[9]

The distinction between Ibn Sina's representational power and the sensitive memory appears in LeDoux's summary as follows:

> The brain mechanisms through which memories of the emotional significance of stimuli are registered, stored, and retrieved are different from the mechanisms through which cognitive memories of the same stimuli are processed. Damage to the former mechanisms prevents a stimulus with a learned emotional meaning from eliciting emotional reactions in us, whereas damage to the latter mechanism interferes with our ability to remember where we saw the

---

9. LeDoux, *The Emotional Brain*, 69.

stimulus, why we were there, and who we were there with.[10]

Thomas emphasised that the estimative power is capable of recognising intentions. Leslie Brothers has contributed to an understanding of the neural basis for such recognition in both humans and animals. Humans and other social animals come equipped with neural systems that predispose them to pick out faces. The amygdala has been shown to be necessary for interpreting facial expressions, direction of gaze, and tone of voice. Brothers has shown that neurons in the same region are responsive to the sight of hands and leg motions typical of walking. Thus, while there are no individual grandmother neurons predisposed to fire in the presence of a particular individual, there are neurons whose function is to respond to visual stimuli that indicate the intentions of other agents.[11]

The foregoing is a sketchy sample of the ways neuroscientists are studying the neural bases of the capacities once attributed to the soul. Of course no amount of empirical evidence can ever disprove the philosophical doctrine of the soul; it can only show that we no longer have need of such a concept.

4. Defining nonreductive physicalism

There are two routes by which to arrive at a physicalist account of human beings. One is to begin with dualism, say, of a Cartesian sort, and then subtract the mind or soul. John Searle has argued persuasively against this move.[12] The other route begins with science. We recognise a certain 'layered'

---

10. *Ibid.*
11. See Brothers, *Friday's Footprint;* for a summary see 'A Neuroscientific Perspective on Human Sociality', in Russell et al, *Neuroscience and the Person,* 67-74.
12. John R Searle, *The Rediscovery of the Mind* (Cambridge, MA: MIT Press, 1992).

feature of reality: subatomic particles at the lowest level combine in increasingly complex structures to give us the features of the world known to chemists, and these in turn combine into incredibly complex organisations to give us biological organisms.

The version of physicalism I espouse denies the complete reducibility of the biological level to that of chemistry and physics. I argue that just as life appears as a result of complex organisation, so too sentience and consciousness appear as nonreducible products of biological organisation.[13] To conceive of how it is possible to get 'mind' out of matter one needs to appreciate not only the development from inorganic to organic, but also from mere homeostasis, through goal-directedness, information processing, goal evaluation, consciousness, and sociality to self-consciousness.

There are a variety of benefits in approaching physicalism scientifically rather than through a reaction against Cartesianism. As Searle has pointed out, it frees one from the (apparent) necessity of attempting to deny or define away obvious facts of experience—such as the fact that we're conscious. Another benefit is this. Arguments against the reducibility of the mental to the physical can draw upon parallel arguments against reductionism in other scientific domains.

The concept of *supervenience* is now used extensively in philosophy of mind. The claim is that, in contrast to earlier mind-brain *identity* theses, it allows for a purely physicalist account of the human person without entailing the

---

13. This is a view long espoused by Arthur R Peacocke and Ian G Barbour. For their most recent formulations, see Peacocke, 'The Sound of Sheer Silence: How Does God Communicate with Humanity?'; and Barbour, 'Neuroscience, Artificial Intelligence, and Human Nature: Theological and Philosophical Reflections', both in Russell et al, editors, *Neuroscience and the Person*; and other chapters therein.

explanatory or causal reduction of the mental. In other words, it leaves room for the causal efficacy of the mental.

The concept of supervenience is better conveyed by example than by definition. Its use began with RM Hare's discussion of the relation between moral and descriptive properties. He says:

> ... let us take that characteristic of 'good' which has been called its supervenience. Suppose that we say 'St Francis was a good man'. It is logically impossible to say this and to maintain at the same time that there might have been another man placed in precisely the same circumstances as St Francis, and who behaved in them in exactly the same way, but who differed from St Francis in this respect only, that he was not a good man.[14]

So Hare is pointing out that St Francis's goodness is not an additional property along with his generosity, trust, chastity. Rather, he is good in virtue of having these character traits. These character traits (given proper circumstances) *constitute* his goodness.

While I have argued that typical definitions of 'supervenience' do not allow for the irreducibility of the mental, it is nonetheless a helpful concept for describing the relation of the mental to the physical.[15] So the thesis I propose here is that humans are complex physical organisms whose complexity results in supervening characteristics such as rationality, emotion, morality, and religious responses.

---

14. RM Hare, *The Language of Morals* (New York: Oxford University Press, 1966), 145. Originally published in 1952.
15. See Nancey Murphy, 'Supervenience and the Downward Efficacy of the Mental: A Nonreductive Physicalist Account of Human Action', in Russell, *Neuroscience and the Person*, 147-64.

5. The problem of free will

The central philosophical problem a physicalist has to answer is this: if mental events supervene upon (or are constituted by or realised by) brain events, and if we assume causal closure at the neurobiological level, how can it *not* be the case that all mental events are merely the product of blind neural causes? If this question cannot be answered then it appears that human freedom is in jeopardy and, even worse, that we are completely deceived about the nature and significance of all intellectual processes—they must be governed by physical causes rather than being governed by reason. Here and in my next paper I shall attempt to shed fresh light on these two issues: the problem of free will, and the problem of mental causation—that is, how can mental events *qua* mental make a difference in the world?

The title of this section is somewhat inappropriate in that there is no such thing as *the* problem of free will.[16] Thus, I need to delimit my topic: I shall not consider the problems raised by theological concepts of predestination, divine omnipotence, or divine foreknowledge. Nor shall I tackle what currently goes under the heading of 'the problem of free will' in philosophical circles. Many versions of 'the problem' amount to ingenious attempts to do the impossible—to argue for freedom, given the assumptions that the universe is deterministic and that free will requires the absence of determinism.[17]

---

16. 'Free *will*' is also a poor term. Looking for inner acts of will in addition to one's actions themselves and inquiring whether the willing was free needlessly complicates an already complex problem. Nonetheless, I shall employ this conventional term.

17. For an excellent critique of many of these puzzles, see Daniel C Dennett, *Elbow Room: The Varieties of Free Will Worth Wanting* (Cambridge, MA: MIT Press, 1983). Much of what follows is indebted to Dennett.

Why Christians Should be Physicalists

There are, however, two interesting kinds of threats to human freedom: various versions of environmental determinism and biological determinism, both genetic and neurobiological. It may be argued that neurobiological determinism is the most basic issue. On the one hand, it is widely agreed that there is not enough genetic information in the genome to determine the details of neural wiring in the brain—environment and random growth account for much of it.[18] On the other hand, the kinds of environmental factors we are interested in (such as social pressures) cannot determine our choices without having some impact on our nervous systems. Thus, the threat of neurobiological determinism seems the most important to address.

My approach will not be to attempt to argue that we do or do not have free will, but rather to assume that we do and attempt to show how agency and free choice emerge out of a neural substrate that may be assumed to be (largely) deterministic. This is an approach to philosophical reasoning recommended by Robert Nozick, who urges philosophers to pursue a particular sort of philosophical explanation in which we bring ourselves to see how something we want to believe could be possible.[19] One not only wants but *needs* to begin with the assumption of free will; to argue the contrary would be hopelessly self-stultifying in that it would amount to giving *reasons* for a philosophical position that *denies* the role of reason in human mental processes.

So the question I shall address in my second paper is this: how is it possible for agency and free choice to emerge out of a deterministic neural substrate—how do deterministic

---

18. See, for example, Hugo Lagercrantz, 'The Child's Brain—On Neurogenetic Determinism and Free Will', in NH Gregersen, WB Drees, and U Görman, editors, *The Human Person in Science and Theology* (Edinburgh: T & T Clark, 2000), 65-72.
19. Described and adopted by Dennett, *Elbow Room*, 49.

processes get organised into a system that possesses a degree of freedom?

6. Theological issues

I turn now to theological issues that must be addressed if a physicalist account of human nature is to be reconciled with the systematic theology that was largely developed using dualistic assumptions. The most obvious issue is what happens after death. The preceding historical survey shows the close tie to eschatological issues. Resurrection of the body has been a mere adjunct to a doctrine of the immortality of the soul for centuries. If there is no substantial soul to survive bodily death then what is to be made of doctrines, formalised at the time of the Reformation, specifying that the dead enjoy conscious relation to God prior to the general resurrection?[20] There are at least two options here. Some consider the biblical evidence for an intermediate state to be both scanty and ambiguous, and claim that the entire person simply disintegrates at death to be recreated by God at the general resurrection. Another approach is to question the meaningfulness of a time-line in discussing eschatological issues. That is, we presume that God is, in some sense, 'outside' of time. If those who have died are 'with God' we cannot meaningfully relate their experience to our creaturely history.

The metaphysical makeup of the person is but one aspect of a much broader topic of theological concern, now designated 'Christian anthropology'; one theological task is to trace the consequences of a physicalist account of the person for a variety of issues such as the place of humankind in the rest of nature, the source and nature of human sinfulness, the claim that humans are made in the image of God.

---

20. This teaching was made official for Catholicism by the Fifth Lateran Council in 1513. Calvin's statement in 1542 has settled the issue for many Protestants.

Why Christians Should be Physicalists

Recognition of the centrality of resurrection to Christian teaching, combined with recognition of the continuity of humans with the whole of nature, calls for reconsideration of the scope of God's final transformative act. There is increased motive to agree with theologians such as Wolfhart Pannenberg who argue that the resurrection of Jesus is a foretaste of the transformation awaiting the entire cosmos.[21] Paul hints at this in Romans: 'For the creation waits with eager longing for the revealing of the children of God; for the creation was subjected to futility, not of its own will but by the will of the one who subjected it, in hope that the creation itself will be set free from its bondage to decay and will obtain the freedom of the glory of the children of God' (Rom 8:19-21 NRSV).

Finally, as Nicholas Lash has argued, a doctrine of God is always correlative to anthropology. When the human person is identified with a solitary mind, God tends to be conceived as a *disembodied* mind, as in the case of classical theism. Much of Lash's own writing argues for the recovery of an embodied and social anthropology in order to recapture a more authentic account of religious experience, but also of a thoroughly trinitarian concept of God.[22]

There are equally important issues to be re-examined in related areas of Christian thought. The concept of the soul has played a major role in the history of Christian ethics for centuries, for example, as justification for prohibition of abortion and for differential treatment of animals and humans. Where do these arguments stand with a revised concept of the nature of the person?

---

21. Wolfhart Pannenberg, *Jesus — God and Man* (Philadelphia: Westminster, 1968). See also the essay by Peters in this volume.
22. See, for example, Nicholas Lash, *Easter in Ordinary: Reflections on Human Experience and the Knowledge of God* (Charlottesville, VA: University Press of Virginia, 1988).

The soul has also long been the focus of spiritual direction and pastoral counselling. What becomes of traditional concepts of religious experience if the person is understood to be purely physical? How is God's revelation to humans to be understood if humans are body rather than 'spirit'? There have been reactions in recent years against the asceticism fostered by Platonic dualism as well as against the tendency to distinguish between saving souls and caring for people's physical needs. Feminist writers have been critical of accounts of gender relationships in which a superior rational soul has been associated with the masculine, and a subordinate material body with the feminine. There is much room for development of more holistic approaches to all of these issues.

7. Conclusion

As you can see, I have raised many more questions in this paper than I have answered. I shall indeed attempt to point the direction to some answers regarding reductionism and free will in my next paper. But for now, I hope I have said enough to stimulate some discussion.

How Physicalists Can Avoid Being Reductionists

Nancey Murphy

1. Introduction
In my first paper I argued that recent developments in neuroscience make body-soul (or mind-body) dualism problematic. In short, given what we know about brain function, what is left for a mind or soul to do? I also suggested that dualism is not a part of original Christian teaching. Thus, it may be wise for Christians who have not already done so to begin the task of integrating a physicalist account of the person into their theological systems.

However, not just any physicalist theory is acceptable. I suggested that Christians need to avoid reductionistic views that see humans simply as Cartesian persons *minus* the soul. The important issue, then, is reductionism: if we are purely physical organisms, how can it not be the case that our thoughts and actions are merely the product of blind laws of neurobiology? My question for this paper is how it is possible for agency and free will to emerge out of a deterministic neural substrate.

2. Freedom as obeying the dictates of reason
I shall start with what I think of as an inadequate account of free will, but work towards a more adequate account. Immanuel Kant argued that having free will is a matter of being capable of being moved by reason rather than by natural causes. If this is an adequate account of free will, then the problems of free will and of what philosophers call the problem of mental causation turn out to be one and the same. While I shall argue later that freedom involves more than this, we shall

Nancey Murphy

have come a long way towards *the* goal of this paper if we can first see how reason gets its grip on neural processes. There is a considerable and interesting body of literature on the problem of mental causation.

2.1 The problem of mental causation

Jaegwon Kim's statement of the problem of mental causation is probably the best known. He argues that mental properties will turn out to be reducible to physical properties unless one countenances some sort of downward causation. But such downward efficacy of the mental would suggest an ontological status for the mental that verges on dualism.[1] The problem can be evoked using a simple diagram. See Figure 1.

$$M_1 \overset{?}{\dashrightarrow} M_2$$
$$\$ \qquad\qquad \$$$
$$P_1 \dashrightarrow P_2$$

Figure 1

Here M_1 and M_2 represent mental states or properties; P_1 and P_2 represent physical states or properties. The arrow from P_1 to P_2 represents a causal relation, and the dollar sign represents the supervenience relation.[2] The diagram, then, represents the

---

1. See Jaegwon Kim, 'The Myth of Nonreductive Materialism', in *The Mind-Body Problem,* Richard Warren and Tadeusz Szubka, editors (Oxford: Basil Blackwell, 1994), 242-60.
2. This use of the dollar sign may be my own most significant contribution to this discussion so far, a fitting symbol to represent the supervenience relation for three reasons: it looks like an 'S' for 'supervenience'; it also resembles one of the symbols for 'approximately equal to,' turned on its

How Physicalists Can Avoid Being Reductionists

assumed causal closure at the physical level—that is, every physical event (in this case, the neurobiological event P_2) has a sufficient physical cause. It also represents the thesis that mental events supervene on brain events.

The dilemma for the nonreductive physicalist comes down to this: Mental properties can be taken to have causal efficacy insofar as they supervene on physical properties and those subvenient physical properties are causally efficacious. But if the physical properties are causally efficacious, what causal work is left for the mental properties? We seem to be left with a new version of epiphenomenalism.

Some philosophers are happy with epiphenomenalism, but I claim we should *not* be happy with this result: in short, it seems to rule out any *reasoned* connection between mental states and to replace them with causal connections. So here I intend to sketch out the basics of an argument for the compatibility of reasoned connections at the mental level with causal connections at the neurobiological level. To do so I shall turn, eventually, to the concept of downward causation.

First, let me make it clear that I am reframing Kim's question. Kim speaks in terms of mental and physical *properties* of events: if the physical property is causally sufficient, what is left for the mental property to do? I want to argue that this way of describing the problem misses the crucial issue. The crucial issue is whether the sequence from M_1 to M_2 is a *reasoned* sequence or merely a *causal* sequence. So, for example, you read '5 times 7'. You think '35'. Did that happen because it is *true* that 5 X 7 = 35 or because a causal process in your brain made you think it?

Given that we presuppose the truth of 5 X 7 = 35, that is, that it is *rational* to think '35' when one reads '5 times 7,' we can again reframe the question: How can we reconcile an account in terms of reasons with a physicalist account of the mental

---

side and squashed together; and, finally, it has the advantage of using a key on the keyboard that philosophers have little use for otherwise.

without giving up on the causal closure of the physical? Colin McGinn asks: 'How, for example, does *modus ponens* get its grip on the causal transitions between mental states?'[3] To sum up the problem of mental causation I would rephrase his question as follows: 'How does *modus ponens* get its grip on the causal transitions between *brain* states?'

2.2 Resources from philosophy and neuroscience
A hint about how I shall proceed: Notice that a calculator obeys the laws of physics *and* the laws of arithmetic. This is because it has been built in such a way that its causal processes model arithmetic transformations. Following Fred Dretske, we can say that the calculator has been *structured* in such a way that any token instance of a series of *triggering* causes—pressing the '5' key, the 'times' key, the '7' and the 'equals'—causes the machine to display a '35'.[4]

So for many purposes it is an oversimplification to represent a causal sequence simply as a series of events: $E_1 \rightarrow E_2 \rightarrow E_3$. Instead we need to think of *two* series of events: those leading up to the triggering of the effect as well as those leading up to the condition under which T is able to cause E. Figure 2, adapted from Dretske's diagram, is intended to represent these intersecting strings of triggering and structuring causes:

---

3. Colin McGinn, 'Consciousness and Content', in Ned Block, Owen Flanagan, and Güven Güzeldere, editors, *The Nature of Consciousness: Philosophical Debates* (Cambridge: Cambridge University Press, 1997), 255-307; 305.
4. Fred Dretske, 'Mental Events as Structuring Causes of Behavior', in John Heil and Alfred Mele, editors, *Mental Causation* (Oxford: Clarendon Press, 1995), 121-136; 122-3.

How Physicalists Can Avoid Being Reductionists

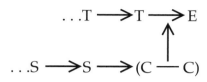

Figure 2

Here the Ts represent a series of triggering causes and the Ss represent a series of structuring causes leading to the ongoing condition C such that T is able to cause the effect E.

So the present question can be rephrased as follows: Are there significant enough analogies between the human brain and a calculator such that we can plausibly assume that the 'wetware' has been structured in such a way that its causal processes model or instantiate rational sequences? The disanalogy, of course, is that the calculator has been intentionally designed by a rational agent. Can we provide a plausible account, based on what we now know of neurobiology, as to how such rational structuring might occur without having to presuppose rational agency? I believe that the answer is yes. But here my account necessarily becomes somewhat speculative due to the tentativeness and incompleteness of neuroscientific explanations.

The physicalist assumption is that a mental event, such as thinking of the number 5 or thinking of Grandma, supervenes on a neural event. As mentioned in my previous paper, it is most plausible to think of such an event as the activation of a cell-assembly. It is in the *training* of such assemblies that we begin to see downward causation. It is better described as downward causation from the environment to the brain rather than mental causation, but insofar as intentionality or reference is an essential ingredient in rationality we have here the beginnings of an account of the rational *structuring* of the brain. Before pursuing this line of thought, however, we need to explore the concept of downward causation.

Nancey Murphy

2.3 Excursus: Defining downward causation

There has been a developing literature on downward or top-down causation over the past forty years. Roger Sperry, who has done more than anyone to promote the concept of top-down causation in the field of psychology, sometimes speaks of the properties of the higher-level entity or system *overpowering* the causal forces of the component entities.[5] However, elsewhere in his writings Sperry refers to Donald Campbell's account of downward causation. Here there is no talk of overpowering lower-level causal processes but, instead, a thoroughly non-mysterious account of a larger system of causal factors having a selective effect on lower-level entities and their causal effects.

Campbell's example is the role of natural selection in producing the remarkably efficient jaw structures of worker termites and ants. He points out that the hinge surfaces and the muscle attachments agree with Archimedes' laws of levers, that is, with macromechanics.

> This is a kind of conformity to physics, but a different kind than is involved in the molecular, atomic, strong and weak coupling processes underlying the formation of the particular proteins of the muscle and shell of which the system is constructed. The laws of levers are one part of the complex selective system operating at the level of whole organisms. Selection at that level has optimised viability, and has thus optimised the form of parts of organisms, for the worker termite and ant and for their solitary ancestors. We need the laws of levers, *and organism-level selection* . . . to explain the particular distribution of proteins

---

5. Roger W Sperry, *Science and Moral Priority: Merging Mind, Brain, and Human Values* (New York: Columbia University Press, 1983), 117.

found in the jaw and *hence* the DNA templates guiding their production.[6]

Downward causation, then, is a matter of the laws of the higher-level selective system determining in part the distribution of lower-level events and substances. 'Description of an intermediate-level phenomenon', he says, 'is not completed by describing its possibility and implementation in lower-level terms. Its presence, prevalence or distribution (all needed for a complete explanation of biological phenomena) will often require reference to laws at a higher level of organisation as well'.[7]

Campbell uses the term 'downward causation' reluctantly. If it is causation, he says, 'it is the back-handed variety of natural selection and cybernetics, causation by a selective system which edits the products of direct physical causation'.[8]

We can represent the bottom-up aspect of the causation as in Figure 3:

jaw structure
$
DNA⟶ protein formation

Figure 3

---

6. Donald T Campbell, '"Downward Causation" in Hierarchically Organised Biological Systems', in FJ Ayala and T Dobzhansky, editors, *Studies in the Philosophy of Biology: Reduction and Related Problems* (Berkeley and Los Angeles: University of California Press, 1974), 179-186; 181.
7. *Ibid*, 180.
8. *Ibid*, 180-81.

That is, the information encoded in the DNA contributes to the production of certain proteins upon which the structure of the termite jaw supervenes. This is micro-physical or bottom-up causation.

However, to represent the top-down aspect of causation, we need a more complex diagram, as in Figure 4, representing feedback from the environment, E. Here the dashed lines represent the top-down aspects, solid lines represent bottom-up causation.[9]

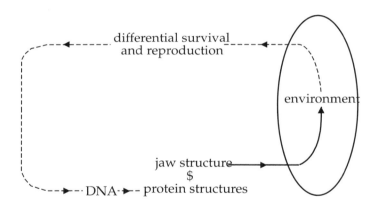

Figure 4

---

[9]. The most helpful recent account of downward causation is that provided by Robert Van Gulick. See, 'Who's in Charge Here? And Who's Doing All the Work?' in Heil and Mele, editors, *Mental Causation*, 233-256.

How Physcialists Can Avoid Being Reductionists

2.4 Downward causation of neural structure

Let us return to the question of how the brain becomes *structured* in such a way that its causal processes realise rational processes. Many theories of brain function rely on 'neural Darwinism'.[10] That is, the answer to the question of how neural nets or cell assemblies form is by a process of random growth of dendrites and synaptic connections, followed by selective reinforcement of connections that turn out to be useful. Useful connections (such as the connection between the 'grandmother' assembly and the 'cookies' assembly) remain strong, while unused connections, (say, between 'grandmother' and 'frogs') weaken or die off. In this way, neural connections that model relations of various sorts in the world come to be selected.

A central claim of my paper, then, is that downward causation, in the sense of environmental selection of neural connections and tuning of synaptic weights, provides a plausible account of how the brain becomes structured to perform rational operations. The larger system—which is the brain in the body interacting with its environment—selects which causal *pathways* will be activated.

So far we have an example of a weak form of rationality—presumably it *is* more rational to think of cookies than of frogs in association with thoughts of one's grandmother. Here the connections among things in the world come to be modelled by connections among cell assemblies in the brain. When this happens, free association is replaced by 'rational' trains of thought.

We can build from this to consider more interesting forms of reasoning. If interaction with the physical world structures the brain in its image, so does interaction with the social world, with its structures and conventions. Consider the social environment of the primary school classroom and the set of conventions we call arithmetic. How do the brains of children

---

10. See, for instance, Gerald M Edelman, *Bright Air, Brilliant Fire: On the Matter of the Mind* (New York: Harper Collins, 1992).

Nancey Murphy

come to be structured so that neurobiological causal processes realise rational operations? Let us speculate about rote learning of multiplication tables. We can imagine that upon hearing the teacher say '5 X 7', neural assemblies are activated and, at first, activation spreads widely and randomly–activating a variety of other assemblies: for example, those subserving thoughts of, '57', 'Times Square', '30', '35', '75'. But feedback from the environment selectively reinforces one connection, while lack of reinforcement weakens all the others. We can picture this process by means of a diagram formally identical to the one I used to represent Campbell's account of downward causation.

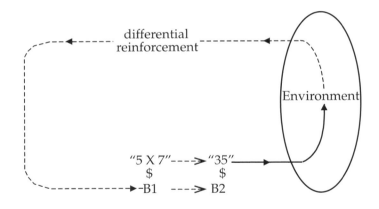

Figure 5

Here the thoughts of '5 X 7' and of '35' are pictured as supervening on two brain states (that is, activation of cell assemblies) labelled B_1 and B_2. Over time, feedback from the social environment results in a strong connection between B_1 and B_2 such that B_1 regularly causes B_2.

How Physcialists Can Avoid Being Reductionists

Let me emphasise that the foregoing is not intended to be a realistic cognitive-science account of the actual learning of arithmetic. In addition, it begs all of the questions pertaining to the foundations of mathematics. It is simply intended to show that downward causation in the form of environmental selection among neural connections provides a plausible explanation of how rational connections could become instantiated in or realised by causal pathways in the brain.

So if my rephrasing of the problem of mental causation is satisfactory, we have here the basis for solving the problem. That is, I claimed that the issue is not the causal powers of the mental properties of events (as Kim says), but rather it is to explain how an account of a sequence of mental events ordered in terms of *reasons* can be reconciled with an account of those same events connected by neurobiological causes.

Note that while we have made sense in neurobiological terms of an agent being moved by reason, we have also shown that Kant was wrong to contrast this with being moved by natural causes. The important conclusion of this subsection is that certain brain processes are simultaneously free (in Kant's sense) and causally determined.

3. Freedom to choose

Kant's conception of freedom as the ability to act in accordance with reason is an unusually narrow analysis of the meaning of free will. In this section and the following I intend to build from the foregoing work on top-down mental causation to a broader conception of freedom. We shall have to ask in the end if the result is in fact an adequate account of free will.

One obvious problem with Kant's understanding is that so few of the decisions we make can be determined solely by reason; life would be much easier if it were possible, when facing a decision, to know in advance *the* optimal solution. As an extreme case, consider the medieval philosophical fiction of Buridan's ass, starving to death between two equidistant and equivalent piles of hay. This thought experiment was designed to raise the problem of reasoned choice in the absence of sufficient reason to choose. Nicholas Rescher states that there is almost no discussion in current literature of the logical issues involved in resolving this problem.[11]

Current neuroscience, though, may have solved the problem. Neuro-ethologists have shown that organisms as primitive as fruit flies and even bacteria exhibit 'initiating activity'. For instance, the bacterium Eschericia coli has a motor that produces 'random' change in the direction it swims. It is able to move to more suitable environments by means of a *delay* in the next change of direction if the milieu improves.[12] Fruit flies exhibit similar periodic 'random' changes in direction, and it has been possible to control their environments sufficiently to show that these movements are internally generated rather than responsive to external stimuli.[13] These self-initiated 'random'[14]

---

11. Nicholas Rescher, 'Jean Buridan', in Paul Edwards, editor, *The Encyclopedia of Philosophy* (New York and London: Macmillan, 1967), Volume 1, 428. Rescher notes that Buridan's opponents were inspired by a similar problem conceived by Al Ghazali involving equally desirable dates (from date palms, not singles' clubs).
12. I suspect that *inhibition* of otherwise spontaneous action is an important aspect of the exercise of free will in the case of humans but I cannot pursue this here.
13. Martin Heisenberg, 'Initiating Activity and the Ability to Act Arbitrarily in Animals', translation by Beatrix Schieffer of 'Initiale Aktivä und Wilkürverhalten bei Tieren', *Naturwissenschaften* 1983 70:70-78.
14. I enclose 'random' in quotation marks because it is not clear whether this is an appropriate term for this sort of event. 'Random' is usually defined as uncaused. These movements are *not* uncaused—the nervous system produces them. However, they *appear* to be random in that they lack any discernible pattern and are not responsive to changes in the environment. Heisenberg attributes the neural basis of these behaviours to spontaneous activity in brain cells: nerve cells that without apparent causes give off rhythmic or non-rhythmic potentials have been found in all types of brains. Are these events truly random (in the sense attributed to certain quantum events) or are there 'hidden variables'?

How Physcialists Can Avoid Being Reductionists

changes in behaviour provide an optimal solution to the problem of survival in an environment too complex to be met entirely with instinctive behaviours.[15] The organism's behaviour thus explores all possibilities in its possibility space (*Verhaltensfreiraum* in Heisenberg's text) and feedback selects the responses that further the organism's goals. Warren Brown (in personal communication) concludes that all organisms with more than the most rudimentary nervous systems have a 'chooser'—a program to produce a multiplicity of behavioural plans that may or may not be used. This being the case, Buridan's ass will not starve; nature has designed it to choose one pile of hay or the other despite the lack of any adequate *reason* to do so. The following diagram (from Donald MacKay) represents a system capable of initiating actions in pursuit of a goal.

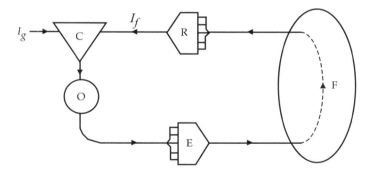

Figure 6

---

15. Dennett, *Elbow Room*, 66-73.

Here F is the field of operation, R is a set of receptors (sense organs, for instance), E is the set of effectors (the fly's wings, for example), O is an organising system that produces responses, and C is a comparator, whose role is to compare information coming from the field of operation (I_f) with information about the goal state (I_g).[16] This diagram is designed to represent any mechanism that pursues a particular goal actively by correcting, automatically, conditions that lead away from the goal and helping trends towards the goal. It can as well represent a thermostat as an organism. I am supposing here only a very primitive organising system, which merely initiates 'random' activity. I could have used this diagram as well to represent the feedback from the environment that shapes the student's learning of multiplication tables. So this is again formally identical to Campbell's top-down causation from the environment.

Notice that another capacity with which nature has endowed even simple organisms is the capacity to change or choose among salient goals. If this were not the case, animal behaviour would not be flexible enough to shift from, say, pursuing a drink of water to escape from a predator. The following figure represents a goal-directed system with the capacity to reset its own goals.

---

16. The diagram is from Donald MacKay, *Behind the Eye* (Oxford: Blackwell, 1991), 42-3.

How Physcialists Can Avoid Being Reductionists

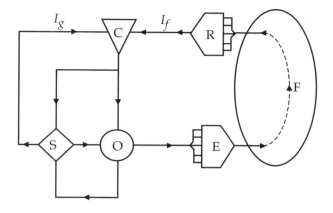

Figure 7

In this figure SS represents a supervisory system that has the power to change the target settings (I_g) so as to settle for a new target because of some unacceptable mismatch being signalled to it from the comparator.[17]

In one of Heisenberg's articles he speaks of 'free will' in the fly.[18] He notes that the spontaneous initiating activity he discusses is *not* an adequate model for human free will. Yet the capacity for such activity is an important prerequisite for the evolution of genuine free will.

4. Free will as self-determination

The foregoing section suggests that although we do not attribute free will to animals there is much to learn about our own capabilities by examining simpler forms of life. What we have seen so far is that organisms are goal-directed systems and

---

17. MacKay, *Behind the Eye*, 50-1.
18. Martin Heisenberg, 'Voluntariness (Wilkürfähigkeit) and the General Organization of Behavior', in RJ Greenspan and CP Kyriacou, editors, *Flexibility and Constraint in Behavioral Systems* (New York: John Wiley and Sons, Ltd, 1994).

that when there is no predetermined behaviour appropriate to the situation they act spontaneously and 'randomly', allowing feedback from the environment to shape their behaviour and to alter the goals they pursue.

What conditions need to be added to this spontaneous but goal-directed activity in order for it to qualify as free will? A central contention of this paper is that 'free' action is best understood as self-determined or self-caused action. Free will is always a matter of degree—no agent ever acts entirely independently of pre-existing biological drives or environmental influence (and no one should *want* to be free in this sense, since biology and society have been shaped to foster our survival). The question, then, is how self-determination gradually *emerges* within the dynamic interplay between biologically driven activity and reactions to the environment. I hypothesise that the following cognitive capacities, found almost exclusively in humans, are the necessary prerequisites:

> 1. symbolic language,
> 2. self-awareness and self-transcendence,
> 3. the ability to imagine behavioural scenarios involving one's own future action, and
> 4. the ability to predict the consequences of such actions.[19]

4.1 Language and self-transcendence
Terrence Deacon argues that language is essential for detaching behaviour from immediate, biologically salient stimuli in order to enable the pursuit of higher-order goals. He describes an instructive series of experiments with chimpanzees. A chimp is given the opportunity to choose between two unequal piles of

---

19. Here I am closely following Warren S Brown, 'A Neurocognitive Perspective on Free Will', *Center for Theology and the Natural Sciences Bulletin*, 19, 1 (Winter 1999): 22-29.

How Physicalists Can Avoid Being Reductionists

candy; it always chooses the bigger one. Then the situation is made more complicated: the chimp chooses, but the experimenter gives the chosen pile to a second chimp and the first ends up with the smaller one. Children over the age of two catch on quickly and choose the smaller pile. But chimps have a very hard time catching on; they watch in agitated dismay, over and over, as the larger pile of candy is given away.

Deacon says that the task poses a difficulty for the chimps because the presence of such a salient reward undermines their ability to stand back from the situation and subjugate their desire to the pragmatic context, which requires them to do the opposite of what they would normally do to achieve the same end.

Now the experiment is further complicated. The chimps are taught to associate numbers with the piles of candy. When given the chance to select numbers rather than the piles themselves, they quickly learn to choose the number associated with the smaller pile. Deacon argues that the symbolic representation helps reduce the power of the stimulus to drive behaviour. Thus, he argues that increasing ability to create symbols progressively frees responses from stimulus-driven immediacy. So language is one piece of the solution to the free-will problem. It helps to account for our ability to detach our behaviour from biological drives.[20]

The experiments with the chimps illustrate a second piece to the free-will puzzle. What the chimps in the first phase of the experiment are unable to do is to make their own behaviour, their own cognitive strategy, the object of their attention. This ability to represent to oneself aspects of one's own cognitive processes so as to be able to evaluate them is what I shall call self-transcendence. Dennett follows DR Hofstadter in pointing out that the truly explosive advance in the escape from crude biological determinism comes when the capacity for pattern recognition is turned in upon itself. The creature who is not

---

20. Deacon, *The Symbolic Species*, 413-15.

only sensitive to patterns in its environment but also to patterns in its own reactions to patterns in its environment has taken a major step.[21] Dennett's term for this ability is to 'go meta'—one represents one's representations, reacts to one's reactions. 'The power to iterate one's powers in this way, to apply whatever tricks one has to one's existing tricks, is a well-recognised breakthrough in many domains: a cascade of processes leading from stupid to sophisticated activity'.[22]

MacKay has a diagram that represents such a system.

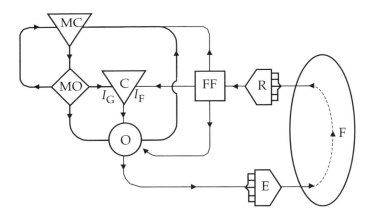

Figure 8

Here the supervisory system of figure 7 is divided into two components: a meta-comparator and a meta-organising system. The MC receives information from the environment by means of a feed-forward system. The MC is capable of recognising not only a mismatch between I_f and I_g but also a mismatch between

---

21. Dennett, *Elbow Room*, 29; referring to DR Hofstadter, 'Can Creativity Be Mechanized?' *Scientific American*, 247 (September 1982): 18-34.
22. *Ibid*.

its own activities and its goals. MC sends information to MO, which in turn directs O to adopt a different strategy.

4.2 Behavioral scenarios and their consequences
The combination of sophisticated language and the capacity for self-transcendence contribute to another dimension of human self-determination, the ability to run mental scenarios involving one's possible future behaviour and to predict their likely consequences. This frees one from the need to implement the behaviour in order to receive environmental feedback. Warren Brown writes:

> Consideration of potential future behavioral scenarios allows for these scenarios to be evaluated with respect to the desirability of the imagined outcomes. The evaluation would be of the kind 'good to enact' or 'bad to enact'. Having been evaluated, the behavioral scenario creates a memory trace that affects the future probabilities of expressed behavior. . . .[23]

Can we put these four factors together now to formulate an account of a free and morally responsible decision?

We come into the world as goal-directed and spontaneously active systems. Conflict between goals that cannot be pursued simultaneously often brings those goals to conscious awareness. Conflicting goals can become prioritised by running scenarios and imagining long-term consequences of various constellations. Language gives us the capacity, in addition, to describe our actual goals in abstract terms and to imagine ourselves pursuing different goals. Thus, the achievement of higher-order abstract goals, such as acting fairly or pleasing God, may become a criterion for ordering lower-level goals.

---

23. Brown, 'A Neurocognitive Perspective on Free Will', 27.

Nancey Murphy

I suggest that a hierarchical ordering of goals of this sort is a prerequisite for responsibility. *When actions are determined in light of such a hierarchy I claim that the person is acting freely.*

Notice that I have described the processes leading up to the 'free' action without invoking any agency—no 'homunculus' within the person's mind making choices. To do so would set up a regress problem: were any of *those* choices free? Rather, I have described a process that could be deterministic all the way through. More likely it is a mixture of deterministic outcomes and of selection among 'randomly' generated alternatives. For example, when faced with a decision there may have been a process of free association resulting in a 'random' selection among all of the possible options that might have come to mind. The highest-level goal might be a product of innate temperament and the various concepts of morality available in the person's milieu.[24] For this reason I think that the question of free will should *not* be taken to hinge on the issue of determinism versus indeterminism.[25] There will always be a host of prior causal events leading up to an act. The important question, I suggest, is whether or not the act can be said to be (largely) determined by the person him- or herself, considered as a goal-directed and self-modifying system. I insert the qualification 'largely' because a human action that is entirely independent of biological and social causal influences is highly

---

24. It is important here to remember that further iterations of self-transcendence are always possible. As soon as I become suspicious that my moral principles are merely conventional I have made them subject to evaluation and may change them as a result.

25. Another related issue in the philosophical literature turns on the argument that free will entails that the agent could have done otherwise, or might do otherwise in exactly the same circumstances. Dennett has argued persuasively against this analysis of free will on the grounds that identical circumstances never present themselves and thus there is no way to know of any given action whether it was free or not. It is a criterion that can never be applied in practice, and it is exactly for practical reasons that we need to know what free will amounts to.

improbable, and if such acts do occur, they are far from the kinds of actions—such as moral choices—that we care about in arguing for free will.

5. Overview

In this section I have not set out to argue *that* humans have free will, but rather have assumed it to be the case as a condition for the meaningfulness of any scholarly work, this essay included. I have attempted instead to fill in a part of the *explanation* of why it is *not* (always) the case that the laws of neurobiology simply determine human thought and behaviour. My explanation involved the key concept of downward causation, and this is in two ways: downward causation from the environment creating patterns of causal pathways that instantiate or realise rational connections; and downward causation from higher-order evaluative or supervisory systems within the agent's cognitive system that reshapes the agent's goals and strategies for achieving them. This downward mental causation also results in the reshaping of the agent's neural pathways.

An adequate treatment of this issue, of course, would require much more space than I have here. What I hope to have accomplished, though, is to show that a physicalist account of the human person is not irreconcilable with our traditional views of ourselves as free and responsible moral agents, and at the same time to suggest that the familiar old battles fought over free will on *philosophical* terrain might better be resolved by taking the issue to be a problem for cognitive-neuroscientists to solve.

6. Conclusion

The topic of human nature is one of the most fruitful points of contact between science and theology. I have considered only a narrow focus of the dialogue: what Christian theology and neuroscience each have to say about the metaphysical constitution of the human being. I claim that Christian scholars and contemporary scientists can agree on a physicalist account of the person. Nonetheless, there are important philosophical problems still to be solved for physicalists who aim for a *nonreductive* account of human thought and behaviour. I hope to have advanced our understanding of how purely physical beings can be both reasonable and free. There is, of course, much more to be done here.

Much more needs to be said, as well, about theological accommodations of a physicalist anthropology, since the Christian tradition has assumed a dualist account through most of its history. I hope what I have written here will be sufficient to stimulate lively discussions of these issues.

Response to Nancey Murphy

Denis Edwards

Nancey Murphy has proposed the idea that all the human capacities that were once attributed to the soul can now be understood in terms of brain functions. She holds for a physicalist view of the human person, arguing that as life appears as a result of complex organisation, so consciousness appears as a product of biological organisation. But she is a *non-reductive* physicalist. She insists that something occurs with human consciousness that cannot be reduced to chemistry or biology.

She sees mental events as dependent upon physical events, as 'supervenient' upon them, but not as reducible to them. This raises the question of freedom. To what extent are we determined by our neurons? Nancey defends human freedom. She argues that downward causation, in the form of environmental selection of neural connections, provides a plausible account of how rational connections can be realised in causal pathways in the brain. Interactions with the physical world and social world structure the brain. She proposes a second important instance of downward causation that helps to explain free will—a downward causation from higher-order supervisory systems within the brain that can reshape a person's goals and strategies. This downward causation results in the reshaping of neural pathways. This means, she argues, that a physicalist account is reconcilable with the view we have of ourselves as free moral agents.

My fundamental response to this work is one of gratitude for the challenge it offers to my own thinking. I have long been convinced of the danger of the dualisms of Plato and Descartes. And, like many others, I have been coming to see that neuroscience demands that we see our mental capacities as profoundly based in the physical brain. All of this leads me to think that we need to consider the human being as profoundly one. At the same time, I am convinced that the theologian Joseph Ratzinger is right when he says that openness to relationships with God and with others is at the heart of the traditional idea of 'soul'.[1] Soul involves relatedness. It involves being a self-in-relation. It involves being a person. This is far too precious a good to be dismissed. I am glad then, that Nancey works hard to show the importance of the non-reductive nature of the physicalism she endorses. I confess that at this stage I am still struggling with the language of 'non-reductive physicalism', because of its weighing towards the physical. But the alternative of 'dual aspect monism' does not seem satisfactory. The 'emergentist–monist' position of Arthur Peacocke and Philip Clayton is closer to my own, but I am not entirely happy with this language.[2] I have not been able to come up with a viable alternative.

My own theological journey of the last couple of years has been influenced by the work Nancey and her colleagues have been doing on these issues—particularly through the books she coedited, *Whatever Happened to the Soul?* and *Neuroscience and the Person*.[3] I thought it might be helpful for me to add some comments as a theologian working within the Roman Catholic

---

1. Joseph Ratzinger, *Eschatology: Death and Eternal Life* (Washington: Catholic University of America, 1988), 155.
2. See their contributions to *Neuroscience and the Person: Scientific Perspectives on Divine Action*, Robert John Russell, Nancey Murphy, Theo Meyering and Michael Arbib, editors (Berkeley, California: Center for Theology and the Natural Sciences, 1999), 181-247.
3. Warren S Brown, Nancey Murphy and H Newton Malony, *Whatever Happened to the Soul? Scientific and Theological Portraits of Human Nature* (Minneapolis: Fortress Press, 1998); Robert John Russell, Nancey Murphey, Theo Meyering and Michael Arbib, editors *Neuoscience and the Person: Scientific Perspectives on Divine Action* (Berkeley: Center for Theology and the Natural Sciences, 1999).

Response to Nancey Murphy

tradition, which I hope are of interest in an ecumenical gathering. They concern two issues: the teaching of God's creation of the human soul, and the teaching of the soul's survival beyond death.

There were two important comments by popes on evolution in the twentieth century. In 1950, Pope Pius XII described evolution as a serious hypothesis, worthy of in-depth investigation.[4] He warned that evolution should not be adopted as if it were a certain doctrine, and as if one could prescind completely from biblical revelation. But he taught that there was no necessary opposition between evolution and Christian faith. He pointed to the conditions on which this doctrine would be compatible with Christian faith. These included safeguarding the doctrine of original sin and the immediate creation of the spiritual soul by God. Then, much more recently, in 1996, John Paul II returned to these issues in a widely publicized address to the Pontifical Council of Sciences.[5] He concluded that new knowledge has led to the recognition that evolution is *more than a hypothesis*. He recognised that new evidence for evolution has emerged from a variety of independent sources. This amounted to significant support for the theory of evolution by a religious leader and it was welcomed by biologists like Stephen Jay Gould.[6] In this

---

4. This teaching was presented in the Pope's encyclical letter *Humani Generis* (1950).
5. 'The Theory of Evolution and the "Gospel of Life"', *Catholic International*, Volume 8, Number 1 (1997): 14-16. The address was given on 23 October 1996. See *L'Osservatore Romano* (English), 30 October 1996.
6. In Gould's interpretation, the pope takes the stance: 'it has been proven true; we always celebrate nature's factuality, and we look forward to interesting discussions of theological implications'. Gould rejoices to find a church leader offering a positive approach to the issue of biological evolution. He describes it as good news and he recalls the wisdom saying: 'As cold waters to a thirsty soul, so is good news from a far country' (Prov 25:25). Gould declares himself to be personally

document, John Paul II refers to Pius XII's teaching of the immediate creation of the human soul. He calls for a philosophical and theological exploration of evolution, which respects both the 'ontological difference' that marks human beings as spiritual creatures and the 'ultimate meaning of the Creator's designs'.

At first glance the insistence of Pius XII on the immediate creation of the soul might appear to commit the Roman Catholic Church to a view of the soul as something independent of the body, but I would argue that this is not the case. For one thing, Roman Catholic teaching has followed Aquinas in seeing the soul and body as radically inter-related as form and matter.[7] Body and soul are not two natures brought together, but together make *one* nature.[8] But more importantly such official teachings needed to be interpreted according to a hermeneutics that distinguishes between the content of a teaching and the modes of expression that may be based on an intellectual framework of a particular historical period.[9]

So what is the meaning of this teaching? Roman Catholic theologians interpret this teaching as a defense of God's action in the constitution of the human being, above all in the creation of the irreducibly spiritual dimension of the human. This teaching does not commit Catholics to any one way of understanding how God creates human beings. Many Catholic theologians resist the concept of a particular divine *intervention*

---

sceptical about the soul, but he adds that he knows that 'souls represent a subject outside the magisterium of science'. His argument is that science and religion need not be in conflict because their teachings occupy different domains, and their *magisteria* (teaching authorities) are 'nonoverlapping'. Stephen Jay Gould, 'Nonoverlapping Magisteria', *Natural History* 106 (March 1997): 16-22.

7. Cf Council of Vienne (1312): DS902.
8. Cf *Catechism of the Catholic Church*, 365.
9. For a clarification of this distinction see the document issued by the Congregation for the Doctrine of the Faith, *Mysterium Ecclesiae* (24 June 1973).

Response to Nancey Murphy

at the origin of the human species, or of a particular divine *intervention* for each individual person. Such intervention is unsatisfactory in terms of science because it does not fit with what we know about evolutionary development. It is unsatisfactory in terms of theology because it reduces God to being a secondary cause alongside other causes. Theologians have interpreted the teaching of God's creation of the human soul as one creative action that 'works through all the generations of living beings, so that everyone shares in this special but continuous action in the great work of universal evolution'.[10] The creation of each human person as a spiritual being is understood as special and unique. Each human being is created in the divine image. But this occurs through God's *one continuous act of ongoing creation*. Michael Schmaus, along with a great number of contemporary theologians, points out that the human being 'is not a creature composed of two elements but is a single being in whom matter and spirit are essentially united'.[11] He sees the spiritual dimension of human beings emerging from within the evolution of life, and springing from the material universe. But the human spirit is not simply an expression of matter and derived from matter. The emergence of self-conscious spiritual beings is something radically *new*. It needs explanation at the level of theology. It can be explained by the special action of the Creator, which is not to be understood in an interventionist sense, but as part of the process of divine ongoing creation, by which God brings forth what is radically new from within the laws and constraints of nature.[12]

---

10. Zoltan Alszeghi, 'Development in the Doctrinal Formulations of the Church Concerning the Theory of Evolution', *Concilium* 6 (1967), 17.
11. Michael Schmaus, *Dogma 2: God and Creation* (London: Sheed and Ward, 1969), 135.
12. *Ibid*, 135-143. For Karl Rahner's views see his 'Evolution: II. Theological', in Karl Rahner, editor, *Encyclopedia of Theology: A*

Denis Edwards

What theologians are suggesting is that God should not be thought of creating individual human beings through a series of interventions, but as creating in one divine act that embraces the whole process. It is this one divine act that enables what is radically new to emerge in creation. Above all it enables the emergence of self-conscious and spiritual human beings. Each of them is created in radical uniqueness in the image of God. Each of them is invited into a unique interpersonal relationship with the triune God in the gift of grace. Each of them is destined for eternal life, which is a participation in the divine perichoretic life of friendship beyond comprehension. The creation of each spiritual being is individual, unique and personal, but can be thought of as brought about through God's one divine action of continuous creation. I find that this view can be brought into dialogue with the kind of view of the human that emerges from Nancey Murphy's analysis.

The second area of doctrine that comes to mind is the teaching of the Fifth Lateran Council (1513) that the human soul is immortal.[13] This has been understood in terms of a separation of body and soul at death, with the soul living on until it is reunited with the body in the final resurrection. This is the theology of the 'separated soul' and the 'intermediate state'.

A number of Protestant theologians have rejected the notion of an intermediate state, and there has been a debate over this issue in Roman Catholic theology. Karl Rahner considers that the doctrinal tradition of the intermediate state and the separated soul can be considered as a stage in the history of theology. It was an intellectual construct meant to safeguard two biblical truths, the individual entry into eternal life at death and the promise of general resurrection.

Rahner himself is convinced that while it is appropriate to distinguish body and soul, they form an inseparable unity in

---

Concise Sacramentum Mundi (London: Burns and Oates, 1975), 487-8.

13. DS 1440. See also the teaching of Benedict XII in *Benedictus Deus* (1336) in DS 1000-2.

Response to Nancey Murphy

the human being. He has always been convinced that there is no such thing as a separated soul. In his earlier work, he tried to address this issue with the idea that at death the finite human spirit is linked in a new bodily way with the matter of the universe. At the death the person does not become acosmic but pancosmic. But later he came to accept the idea of a 'resurrection in death'. Gisbert Greshake had introduced this idea in 1969, arguing that the whole person dies, and God raises the whole person in the moment of death. Greshake and Rahner argue that there is no disembodied soul. Nor is there an interval in which the person ceases to exist before being recreated by God, as there is in some recent Protestant theology. Rahner is well aware that the identity of the human being beyond death cannot be established by the idea of some enduring fragment of the metabolic earthly body. He sees identity as established by the free spiritual subject, which is called the soul. This transformed soul 'informs' the glorified body.[14] Resurrection is the fulfilment of the one free spiritual subject and the consummation of the person's history of freedom. It is the fulfilment of the whole person. Joseph Ratzinger has contested the idea of resurrection in death. He is concerned to defend the idea of a communal resurrection that involves the whole of cosmic and human history.[15] The debate goes on, and recently Bernard Prusak has resurrected the idea of the intermediate state without the separation of the soul. He proposed a distinction between the individual resurrection in death from the general resurrection at the end of history and time.[16]

---

14. Karl Rahner, 'The Intermediate State', *Theological Investigations* XVII, 10-124.
15. Joseph Ratzinger, *Eschatology*, 181-2. These concerns were echoed in the International Theological Commission's document 'Some Current Questions in Eschatology' issued in 1992.
16. Bernard P Prusak, 'Bodily Resurrection in Catholic Perspective', *Theological Studies* 61 (March 2000): 64-105.

Denis Edwards

I hope I have said enough to show that there is a lively debate in Roman Catholic theology over these issues. What is common to many theologians in the Catholic tradition is a rejection of the idea that there can be any such thing as a separated body-free soul. And all agree that the human being is fundamentally one. I believe that there is a great deal of room here for further dialogue with the kinds of positions that Nancey has argued.

Reflections on Artificial Intelligence, Emergence and Agency

Adrian Wyard

1. Introduction

In this paper I shall attempt to look into the future and identify what I see as the most interesting questions that advances in information technology (specifically artificial intelligence) will bring before us. This is no easy task. As Neils Bohr has said, 'prediction is very hard, especially regarding the future'.

But first a word or two is needed about the perspective I bring to this task. Until 1995 I was a software designer for a large company, and on entering the multidisplinary education world in 1996 I had few expectations that my work designing applications would have much relevance to scholarly work in this new field. My expertise was in engineering rather than pure computer science research, and I did not see that artificial intelligence (AI) or information technology (IT) prompted any *entirely* new ethical or religious questions or ethical dilemmas. It seemed to me that they just warmed over age-old problems (free will, other minds, personhood, et al). However, I do now believe that AI and IT offer valuable new vantage points from which to view some of these perennial problems. At a minimum these new perspectives can be useful when teaching about several philosophical/ethical challenges, and may even ultimately lead to progress with them.

I must also acknowledge that many people have written authoritatively along similar lines before me. My intent is to be spare when covering what I believe to be well-trodden ground, and go deeper where I think I have something original to say.

2. Ontological classification by common sense?

I believe that AI research causes us to reflect on some of the most basic philosophical questions, including: What is out there in the world? What is real?

Any science—any rational discourse—involves first dividing up the world of our experience into categories, into objects with specific properties that we can then investigate and describe to each other. This is an intuitive process and on most days is so obvious as not to need mentioning. There is no human alive who would not see the world as divisible into plants, animals, other humans, water, rocks, sky, homes, tools and so forth.

We also tend to intuitively arrange these groupings into various hierarchies. One common hierarchy is the degree of autonomy or *agency* objects exhibit. The more increasingly *alive/animated* an object seems, the more attention it usually commands: a lion is more interesting than a worm, which is more interesting than a flower, which is in turn more interesting than a pebble. Of course, dividing up the world this way makes a lot of sense. We pay more attention to the lion, since it is clearly an *agent* in the world, whose actions we cannot predict. It might, for example, try to make us its lunch. In contrast, the worm is an agent of a minimal sort; but has no effects that are of concern to us. The pebble just sits there and can be safely ignored.

The degree of agency that an object exhibits is intimately connected to our ethical relationship to the object as well. If a human steals my lunch, they—their body—is the responsible agent that stole, and this is *where* I should seek recompense. If I trip over a tree root, I can hardly blame the tree since it is not an agent. Trying to get justice from the tree would be a foolish project. Interestingly, if a human trains a dog to steal my food, it is unclear whether we should trace blame back to the trainer, or distribute it across both trainer and trainee . . . Notions of individual identity, personal responsibility and transformation are basic to religious thinking too, and these all hinge on agency.

Reflections on Artificial Intelligence, Emergence and Agency

The degree of agency is also often related to the value we place on an object. To destroy a worm, sheep, or human, is a progressively more serious act, tracking (among other things) the amount of agency each exhibit. Ethical dilemmas emerge when we consider cases where capacity for agency no longer correlates to perceived value. A PVS (persistent vegetative state) patient *looks* very much like other people who are agents in the world, but they are not capable of the same behaviours. If they are to be afforded full human status, we will have to look somewhere other than their capacity for agency.

3. Digital AI breaks out of the categories

So far so good, but today's digital AI systems dramatically fail to conform to this categorisation. Historically, if something roared at us, or *did* much at all, it was safe to assume it was an agent. The more human-like the behaviour, the more agency we were encountering. As already mentioned, the degree of humanness and agency in turn cues both how much we value the object, and how we plan to relate to it. As Brian Cantwell-Smith has noted, until recently if anything spoke to us, we could hope to take it home for dinner, but these days anything from our cars, to computers, or artificial intelligences might try and strike up a conversation with us.

For the near term, intelligent devices will be able to exhibit all sorts of behaviours that make them *look* as though they are agents, while we can be assured that by many measures they are in fact *not* agents. Chess playing has traditionally been associated with high intelligence, and there is no doubt that IBM's chess-playing system Deep Blue that beat the world champion Kasparov in 1997 was a magnificent technical achievement. As might have been expected, the media reported that a brave new era of artificial intelligence had begun. But the researchers themselves were far more gaurded. According to Chung-Jen Tan, 'This chess project is not AI', and Joseph Hoane, 'The techniques that tried to mimic human judgment failed miserably. We still do not know how to do that at all.'

4. Proposal: Digital (deterministic) systems cannot be agents

I can make such a categorical statement about the lack of capacity for agency in Deep Blue and comparable systems because they are entirely, or at core, digital. Since they are digital, we know that they are fully deterministic; there is no behaviour they can exhibit that is not reliably traceable, after the fact, to some external source. The source could be the programmer, or a stimulus from the environment, or a combination, but is always external to the digital system. If such a device were to be involved in an ethical situation—say a robot dog was programmed to find and steal my lunch—then I think it is clear that the human programmer is the agent that stole my lunch, and it is with that person I should seek justice, not with the robot. In a sense, while the robot *looks* to be an agent distinct from the programmer, the robot can be seen as just an extension of the human programmer—it does not look like it, but the two are a single agent.

5. What can be an agent?

The description given above assumes that digital systems cannot be agents because they are deterministic, and that humans are potential agents because they non-deterministic. I should say a word or two about what leads me to this assumption. I cannot offer a sophisticated metaphysics that shows fundamental determinism to be false, or that proves humans to be a special class of entirely free agents; rather, I am just appealing to epistemic brevity. *Something* has to be an agent if we are to carry on conversations.[1] I believe the most useful (brief, efficient) descriptions of situations that include conscious human persons can legitimately include them as potential

---

1. With no agents, it seems to me we are left with a full-blown deterministic, atemporal, block-universe description, which I really do not know how to deal with.

agents simply because I do not see how to usefully trace some of the effects of people anywhere else.[2]

Having said this, I do not mean to suggest that distinguishing among agents is always clear-cut, or that humans are not subject to influence from agents outside themselves. Certainly, when analysing a situation involving a human person, we could conceivably properly trace causes to a subperson level, perhaps to the level of a biological or genetic function where there is no real influence from the level of consciousness. Similarly, the best place to locate agency might sometimes be at the multiple-person level—where group dynamics are the proper root cause of effects.

For real-life scenarios involving people, we should probably expect to find a plethora of inter-related agents at various scales, from atomic, to genetic, to person, to group, to ecosystem. But if we were to rank each of them by the quantity of effects traceable to each, I think the conscious human person would invariably be at the top of the list, and any digital systems present would often be near the bottom, close to the rocks.

The presence of intelligent digital systems in our world, and our predisposition to interpret agent-like behaviours as real agency highlights the importance of locating what might be called *internal* agency. By this I mean the minimally complex grouping of parts and subparts where the effect we are looking for can be reasonably said to originate. Such a scheme would not be tripped up by the appearance of a freely roaming robot dog—it would group the programmer and the dog machinery as a single agent with the capacity for lunch-stealing.

However, I should admit that this approach is immensely subjective. If we start off looking for the cause of bacterium-

---

2. I take causation as a helpful guide to ontological classification. Since I see 'mind' as a potential agent, I count it as 'real.' Quite how the mind is real is a separate question, but I am quite convinced by the non-reductive physicalist accounts of Murphy, Brown, *et al*.

style behaviour in a human subject, we will find our agent long before reaching the conscious human level. Similarly, if we were to use this technique to determine if some future[3] A I systems deserve to be called agents, we should be careful not to look for overly anthropocentric effects.

It can also be argued that such a search for internal agency can only be successful while we are ignorant as to the functioning of lower level parts of complex systems, as is currently the case with the human brain. (Certainly we cannot 'trace' behaviours after the fact as we can with digital systems.) Perhaps with more knowledge we will find that human agency at the conscious level is just an illusion, and that it should properly be located at lower levels, perhaps eventually sliding all the way down to nothing but physics and chemistry. If this were the case, the difference between digital AI and ourselves would be far less than I am claiming. This is a possibility, but two related factors lead me to doubt it:

First, the basic unit of functioning is quite different: Compared to the extremely reliable electronic-digital interactions in a circuit, the biochemical interactions in a cell are quite fluid (literally!)—reliability is sometimes achieved only by redundancy. Some systems are tightly coupled, but some are not—there is built-in wiggle-room. Second, the sheer number of interactions of parts makes us susceptible to complexity effects, even if the parts are largely deterministic.

So, it seems to me that such an approach, while limited, has some utility even if it is only a slightly formal expression of human common sense. One thing it does do is show a sharp division between today's digital AI and human persons, even when AI systems exhibit compelling human-like behaviours.

6. Defining the AI task

Does this mean that all the hard AI questions are in fact easy since there will never be an *agent* there? Not at all. There are

---

3. Non-deterministic.

many approaches to building intelligent systems that do not rely solely on digital foundations, for example, the embodied sysems work at the MIT AI Laboratory in the United States. These cannot be dismissed out of hand as non-agents. Instead we need to look at the physical-mechanical parts they are made from, and decide what capacities they might have. Due to their simplicity, they would still be near the bottom of our ranked list of agents, but above purely software-based digital systems.

So what are the prospects for artificial intelligence, perhaps as envisaged as HAL 9000 in *2001: A Space Odyssey*? There is still much disagreement on this. The problem is threefold. First, we do not have agreement on a test that would signify the existence of 'real' intelligence. Second, we consider ourselves to be intelligent, so testing for human-like behaviour tends to take centre stage. But, is such an anthropocentric perspective appropriate? Many would argue that cyber-intelligence will have its own specific character that may be quite distinct from our own. Third, we do not yet agree on what it is that enables us to be intelligent. There are several options:

- Dualism or Trichotomism: Here the human person is understood as made up of body and a mind that is distinct from the body, or a body, mind, and soul respectively. If this view is correct, then AI as currently understood will never succeed because we do not know how to insert a mind or soul into the machine.

- Non-Reductive Physicalism (or Emergent Monism): Here the person is made up of a physical body, with no separate mind or soul added. Rather, the mind is considered a causally effective emergent property of the constituent parts of the brain/body. Such a description does not rule out AI in principle, but since digital systems eradicate emergent properties by their very design, if this view of human personhood is correct, digital systems are not a good foundation on which to build something intended to be a mind-like agent.

- Reductive Monism: The person is made up of nothing but the physical body. The functioning of the whole body, or any part, can be expressed in terms of (reduced to) biochemical interactions. Since notions of mind and personhood have no meaning at the level of biochemistry, they may therefore be illusory. If this description is correct, then by analogy, approximating the function/complexity of human biochemistry may be all that is needed to create AI.

- Behaviourism: Here human persons are understood solely in terms of their external behaviours. Other aspects of humanness such as consciousness and intentionality/agency are excluded on philosophical grounds. If AI were to duplicate human behaviours, regardless of how, then a behaviourist perspective would consider it equivalent.

For the purposes of this paper I will not pursue the dualism and trichotomism options that rule out the possibility of AI on metaphysical grounds.

8. On being an agent and kinds of freedom

As we make our way through these issues it would be helpful to have crisp definitions for terms like 'intelligent', 'machine', 'artificial', 'machine intelligence' and so forth. As mentioned above, this is tough task. I shall use the potential for agency as a way to rank and group capacities. At the bottom of such a list would be inanimate matter—a rock, for example. If a person were included in the scenario, they may use the rock for some purpose, and any effect the rock had in our scenario would be traced back to the human agent. I will use the term 'tool' for this kind of object and scenario.

Reflections on Artificial Intelligence, Emergence and Agency

A rock has no internal degrees of freedom—it is solid. As we add complexity to tools they fall into another category—machines. Machines have moving parts, joints, and the parts can be in various configurations. An example of something in the machine category would be a drill, or a loom. Are machines like drills and looms ever agents? This is a tricky question. Agency will come into the equation as we add either complexity (more parts with more degrees of freedom), or certain kinds of freedom, or both. But even at the level of looms and drills we can contrive some thought-provoking scenarios.

Take a revolver for example. If we imagine one bullet is placed in the gun, leaving five chambers empty, and the cylinder is then given a good spin and fired by a human operator, where do we trace the effect of the gun? At first glance, a revolver does not seem very different from a normal pistol, but it is clear we can trace the effect of the regular pistol to the person who pulled the trigger, but can we do so with the revolver? I do not think so. This is because the way that the parts of the revolver are put together apparently lends itself to agency in a way that the construction of a regular pistol does not. At a minimum we have to say that the human operator has less control over the effect of the gun if it is a revolver. It seems we have to locate this *spare* agency somewhere, and the gun is the most obvious place. In other words: The behaviour of a revolver with one bullet is contingent on the final resting position of the cylinder, and this is in turn so difficult to predict, that we may as well call it probabilistic (there is about a 1/6 chance the gun will fire a bullet each time the chamber is spun).

To use another metaphor: there is a significant difference between the behaviour of square pegs in square holes, and round pegs in round holes.[4] If we imagine a round peg loosely placed in a round hole and then jiggle the system, the resting place of the peg is hard to predict—our jiggling may caused it to rotate by an arbitrary amount with respect to the hole.

---

4. For now I will avoid the question of round pegs in square holes.

However, a square peg in a square hole, with the same clearance, and jiggled in the same way, will remain locked in roughly the same orientation with respect to the hole. When we are evaluating a scenario to find possible agents, those made up of 'squarely' connected (pistol-like) parts will allow us to trace causes elsewhere, but those made up of 'roundly' connected (revolver-like) parts will not allow us to do so as easily, and will therefore be candidates for agency.[5]

A brief aside on Newtonian dynamics and determinism in general is appropriate. I do not mean to suggest that huge swaths of the world around us are not subject to exact prediction by simple Newtonian calculations. I also concede that the *deep* question of fundamental indeterminism at the quantum level is still open to question. But for real-life scenarios there seems to be quite sufficient warrant for significant epistemic humility, based if not on the Heisenberg uncertainty principle, then on the sheer complexity of the world. One of my favourite examples of real-life unpredictability is the *falling fork problem*. All of us have knocked a fork off a table before, and Newtonian mechanics will predict the maximum radius of the circle in which we can expect to find the fork, but the exact location and orientation of the fork within that circle is essentially beyond our predictive powers—especially if the floor-surface is hard. Similarly with a roulette wheel: We can safely predict the ball will eventually fall toward the centre of the wheel, but not on which number it will come to rest.

9. From intelligent machines to persons

In recent history we have seen the advent of autonomous machines and intelligent machines. These are distinct from machines like looms in that they can change the configuration of their parts without operator involvement. A steam engine

---

5. Digital systems are the most 'squarely' connected devices available.

Reflections on Artificial Intelligence, Emergence and Agency

will happily continue its power cycle for a while without attention, and a clock will continue ticking.

Intelligent machines have the ability to reconfigure their parts in appropriate response to an input of some kind. A response is considered appropriate/intelligent if it corresponds to what a human would do if they were doing the task. A thermostat is a primitive intelligent machine—it behaves as a person would if they were sensing when a room becomes too cold or too hot.[6] A mechanical or electronic calculator is intelligent because it shows the results of calculations that a human might do, but of course with far greater accuracy and speed than any human.

The pioneers of the computer age realised that pretty much any problem could be broken down into a series of simple calculations (sometimes *very, very many* simple calculations), which led to the intelligent machines that surround us today; the general-purpose computer. Every digital computer in existence today is still directly related to the first systems that could do no more than add small non-decimal numbers, and store them in memory locations.

The next obvious question is this: Can sufficiently powerful computers represent, or duplicate, the intelligence of humans? While I have already stated I see problems with the idea that digital technology will produce something that would *count* as a free person, I see no limit to the theoretical power of computation. Some see the use of digital/numerical systems as unlikely to produce human-like behaviours because they intuit that there is something important about being made up of chemicals—being *squishy*. I am not so sure. It seems to me that any systems, even squishy ones, can be represented digitally just so long as the model is sufficiently accurate.[7]

---

6. The behaviour of the governor in a steam engine corresponds to the notion of too fast (dangerous) and too slow (will stall) and so counts as an intelligent machine.
7. I would argue that even *intrinsically* analogue forms such as the pressure waves that make up music may can be *fully* represented without loss, just

Adrian Wyard

However, what is theoretically possible and what is practical may prove to be dramatically different. While the average desktop PC today has a hard drive sufficiently large to represent each of the ~100 billion neurons in a human brain,[8] brute force representations of the human brain as a simulated neural net require computation power and storage that is far beyond our current capabilities. In fact, by Owen Flanagan's calculations, there is not sufficient matter in the visible universe to create a computer large enough to represent a single human brain state.[9] I can safely say the reality is somewhere between these two!

Based on estimates of the computational capabilities of the brain, Ray Kurzweil predicts that computers will exceed our capacity[10]—and therefore human intelligence—by about 2020. Kevin Warwick and others put the date a little later, around mid-century. While I agree that the computing power of the next several decades will enable amazing things, power alone does not make intelligence. We are still far from understanding the right software approach that will lead to the kind of learning that we see in humans. Furthermore, I do not see intelligence, narrowly defined as mathematical abilities, or chess-playing, or vocabulary/language skills, or general knowledge, as something that would indicate equivalence with human persons. But I have no doubt we will find fruitful software solutions to the problem of representing pseudo-

---

        so long as the representation is of sufficient resolution that the steps/artefacts introduced by the digital model are less than the noise one would expect to hear when experiencing the signal through non-digital means.

8. That is, one bit per neuron on a twelve gigabytes or larger hard drive.
9. Owen Flanagan, *Consciousness Reconsidered* (Cambridge Mass: MIT Press 1992), 35-37
10. Estimated at twenty million billion neural connections per second.

intelligence, and that intelligent machines will transcend this category to become artificial intelligences.[11]

But is there a bridge from ever more powerful conventional digital computation all the way to personhood? My intuition is that this is not the way things will go. For some time the AI research community has been pursuing directions other than conventional 'strong AI', focusing instead on embodied intelligence, or 'situated AI'. Here, the objective is to create robots whose intelligence is a result of their physicality as well as a digital computational component. Rather than a central control program that directs the parts of the robot by maintaining a high-fidelity software model of the robot's state and the world around it, the individual parts of the robot have a degree of autonomy from the rest of the robot, and importantly, this may be implemented with non-digital technology.[12] It is my prediction that around mid-century a fuller understanding of developmental genetics, neurobiology, and the limitations of conventional digital computational models of intelligence will lead us to look to an artificial-biological technology as the best approach for developing human-like intelligence. Whether this will be predominantly biological or computational is not clear.

10. AI ethics and religious experience

The incorporation of biological (that is, 'round') functions into future AI efforts would mean that they would be more convincing agents than purely digital systems, and the attendant ethical questions would become more pressing. Can we turn it off? Can we use it for spare parts? If we were to bring intelligences into the world that appeared to learn, and could be said to be distinct agents with capabilities that seemed similar to humans, then would we afford them personhood? Chances

---

11. I prefer the term 'machine intelligence', but AI is a firmly entrenched term.
12. The Cog and Kizmet projects at MIT are good examples of work in this area.

are, we will first need an intermediate category where the behaviours are not quite as rich as a human person, but nevertheless on a par with children, or intelligent animals. What should we call these? How should we treat them?

It seems that we can take cues from how we relate to the animal kingdom. It is possible that systems will be ranked by their capacity in terms of neuronal (or simulated neuronal) connections. A kiloneuron (10^3 connected neurons) intelligence would be analogous to an insect, and perhaps just as we sometimes exterminate insects we would not consider it a terrible crime to terminate a 1 KN intelligence. A meganeuron (10^6) system would perhaps be similar to some small mammals, and a giganeuron (10^9) system would be similar to a higher mammal such as a dog. Perhaps at this level we would require a license to manage such a system, and to provide it with what counts as care and facilities. As we continue up the scale to teraneuron (10^{12}) systems we approach and then exceed human neuronal capacity. Would it be murder to terminate such a system? Would we allow them to vote, or take on positions of leadership?

These are tough questions, and the number of neurons is little more than a measure of capacity. We know that with humans, the possession of neuronal capacity does not make for intelligence, let alone 'good' behaviour.

And what of capacity for religious life? When an Oxford theologian was asked if he would baptise a robot, he gave what I take to be a very profound and helpful answer: 'If it asked properly'.

Reflections on Artificial Intelligence, Emergence and Agency

Suggested Reading

Brown, Murphy, Maloney, editors, *Whatever Happened to the Soul?* Minneapolis: Fortress Press, 1998.

Kevin Warwick, *In the Mind of the Machine,* Arrow, 1998.

John Puddefoot, *God and the Mind Machine,* SPCK, London, 1996.

Ray Kurzweil, *The Age of Spiritual Machines,* New York: Viking, 1999.

Jennifer Cobb, *Cybergrace: The Search for God in the Digital World,* Crown: New York, 1998.

Part Three

Theological Perspectives

Science, the Laws of Nature and Divine Action

William R Stoeger SJ

Over the past fifteen years we have gradually begun to realise that the issue of divine action is central to the science-theology dialogue. This has certainly been expressed in the Vatican Observatory/Center for Theology and the Natural Sciences workshop series, the proceedings of which always have the subtitle, 'Scientific Perspectives on Divine Action'. And it has become a focal emphasis in many other discussions and treatments as well.

1. Categories of divine action

There are two general categories of divine action which must be distinguished.

First, God's universal creative action in nature; and second, God's special action in history. The first is relatively unproblematic. God's universal creative action is what God 'does' to bring and keep all things in existence and maintain the evident order of reality. It has two complementary aspects: creation from absolutely nothing (*creatio ex nihilo*), which asserts the radical dependence of everything on God, whether or not a temporal beginning occurred; and continuing creation (*creatio continua*), which expresses the necessary conservation of all things in existence and in order by God from moment to moment. In our evolving universe, (*creatio continua*) is also realised through the continuing emergence of new entities, systems, species through the operation of the laws and dynamisms of nature—God continuing to create through its processes, regularities and relationships.

As such there rarely seems to be any intervention on God's part in divine universal creative action, nor any 'God of the gaps', except the ontological gap between something and absolutely nothing.

The second category, God's special action in history, is quite different. It is not universal, but rather is focused in what God 'does' by virtue of initiating, maintaining or responding to God's 'personal' covenantal relationship with persons and believing communities. Examples are the Exodus events, the prophetic utterances and actions addressed to Israel, the incarnation, the resurrection, the sending of the Spirit, miracles, answering prayer. This special action of God always *seems* to involve God's intervention, and the 'violation' of the laws of nature. Despite the distinction we have drawn here, there may be a deep connection between God's special action in history and God's universal creative action, at the level of a thorough understanding of the relationship between God and nature. Thomas Tracy has begun to explore this connection in a recent article.[1]

In speaking about divine action or causality, we must always remember that we are attributing such categories of behaviour to God only analogically—the action and causality which God exercises is significantly different from any other kind of action or causality. That is the reason why we sometimes refer to it as 'primary causality'. It is more fundamental, but also more immanent or interior, than any other, and, though it often employs, it does not need, a medium or a pre-existent potentiality.

---

1. Thomas Tracy, in *Quantum Physics: Scientific Perspectives on Divine Action*, Robert J Russell, John Polkinghorne, and Phillip Clayton, editors (Vatican Observatory Publication and Center for Theology and the Natural Sciences, in press, 2001).

Science, the Laws of Nature and Divine Action

2. Divine action and the laws of nature

How we speak about divine action in nature and in history depends a great deal on how we conceive 'the laws of nature'—for example, whether or not divine action in a given case involves 'intervention'. The American philosopher William Alston has commented:

> ... I believe that divine action at particular times and places is quite consistent with physical laws, even of a deterministic form. Whether this is so or not depends on the right way of thinking of such laws. To suppose that divine intervention in a physical process would involve a violation of physical law depends on thinking of physical laws ... as specifying *unqualifiedly* sufficient conditions for an outcome.[2]

What is the *right* way of thinking about 'the laws of nature'? That is the focus of this paper.

3. The laws of nature

Our knowledge of the laws of nature emerges as one of the principal products of the natural sciences. But what are the natural sciences? How can we characterise them? They are the disciplines oriented towards the detailed qualitative and quantitative modelling and understanding of the principal regularities, processes, structures and interrelationships which we find in the natural world—relying on rigorous, repeatable analysis and experiment or observation. These principal regularities, processes, structures and interrelationships *are* the laws of nature. But, notice, from our definition of the natural sciences, there are really two meanings 'the laws of nature' can

---

2. William P Alston, 'Divine Action, Human Freedom, and the Laws of Nature', in *Quantum Cosmology and the Laws of Nature*, edited by Robert J Russell, Nancey Murphy and C J Isham, editors (Vatican Observatory Publications and CTNS, 1993), 189.

assume. The phrase may mean either 'the regularities, processes, . . . ' as *we* imperfectly and provisionally model them, or describe them—'our laws of nature'. Or it may mean 'the regularities, processes, . . .' as they actually function in reality, as they actually are in themselves. We can easily see that this is an important distinction to make—for 'the laws of nature as they actually function' are much more intricate and extensive and adequate than our models and descriptions of them.

4. 'Our laws of nature'

There are various ways in which we can appreciate the provisionality and the limitations of the laws of nature as we model and describe them—how they may certainly fail to represent the underlying regularities, relationships and processes in nature as they actually function. It is helpful to briefly mention some of these.

First of all, 'our laws of nature' are approximate descriptions of the underlying regularities and constraints—*not* giving the fundamental necessities themselves, only attempting to model or describe them. In doing so they effectively describe *some* of the constitutive relationships which contribute to the existence and identity of entities and sytems, but they never account for all of them, nor do they describe fully those they do treat. Secondly, the status of the theories upon which 'our laws of nature' are based vary a great deal. In some cases, these theories are very well established, while in others they are much more phenomenological or speculative. It is worth noting that, when we speak of the laws of nature as if they are complete and immutable, we are almost always speaking of the underlying regularities or constraints in nature, *not* our descriptions or models of them.

Thirdly, we often encounter the phenomenon of theory change or paradigm shift, for instance in the rise of special and general relativity, in the emergence of quantum theory, or more recently in the importance of complex systems and chaos. They have occurred under the pressure of unexplained phenomena.

Science, the Laws of Nature and Divine Action

These changes or transitions do not always signal the complete abandonment of a theory or model, but often the subsuming of old, well-tested theories (within the realm of their validity) into newer, more adequate theories. For instance, Newtonian physics was subsumed as a special case in relativity, as was classical physics in quantum physics. Typically, correspondence principles relate the older theories to the newer more adequate ones. There is always a range of cases in which the old theories and the newer ones describe and predict the behavior of systems with equal accuracy—and in which the old theory describes it in a simpler, more intuitive or more straightforward way. The new theory almost always relies on a very different paradigm, which explains and describes the behavior of systems as accurately as the old theory in its realm of competence but also describes it more accurately in situations which are beyond the realm of competency of the old theory—for instance, in the case of general relativity, for strong gravitational fields and high velocities (near that of light). In this case, can we really say that the newer theory with the larger scope is always better or 'more true'? Not necessary, argues Charles W Misner in a now famous essay.[3]

There is a fourth limitation of 'our laws of nature'. They always prescind from other aspects of reality than those upon which they focus, and have been formulated based on idealisations. For instance, the laws of mechanics prescind from personal and ethical issues, and are based on considering point masses in simple motion. They leave out any complication or complexities.

Despite their limitations, they do *work,* and accurately describe some of the basic features of physical, chemical and biological reality. But they also leave a lot out—even at the physical, chemical and biological levels. They often do describe

---

3. Charles W Misner, 'Cosmology and Theology', in *Cosmology, History and Theology*, Wolfgang Yourgrau and Allen D Breck, editors (New York and London: Plenum Press, 1977), 75-100 (cf especially 97-99).

and model some constitutive relationships, but fail to account for others.

Even those they do describe and model, for instance gravity, they do so only provisionally and incompletely. The natural sciences almost never yield a complete and thorough understanding which accounts for the phenomenon being as it is. Thus, as I have insisted above, 'our laws of nature' are *descriptive*, but not prescriptive. They cannot locate and express the underlying necessities which give the laws of nature their force. They are simply *models*, not the regularities, processes and relationships themselves.

How accurately do these models represent the laws of nature at they actually function? We can only determine this by experiment and observation—to see, for instance, how well the model predicts outcomes of an experiment, or describes qualitatively and quantitatively the phenomena we observe and their relationships to other phenomena. But this only takes us so far. It is really not possible to determine reliably how accurately our models reflect the relationships, regularities and constraints of reality at the deepest level and their connection with one another. To do this we would need a privileged purchase on the relationship, regularity or process in question, independent of our observations of it, and of our theory about it. That is simply not possible.

What we actually do in the natural sciences is employ what Charles S Peirce refers to as retroduction (abduction), a variation of the hypothetico-deductive method. As Ernan McMullin summarises it we first, on the basis of what we know about the phenomenon or system in question, form a hypothesis concerning the underlying or hidden processes or realities at the root of the phenomenon.[4] On the basis of that

---

4. Cf Ernan McMullin, 'Structural Explanation', *American Philosophical Quarterly*, 15: 139-147 (1978); 'The Shaping of Scientific Rationality', in *Construction and Constraint: The Shaping of Scientific Rationality*, Ernan McMullin, editor (Notre Dame, Indiana: University of Notre

hypothesis, certain consequences should follow—predictions. We test the hypothesis and modify it in light of those tests until we have a refined hypothesis which is confirmed by subsequent tests. If a hypothesis is fruitful and successful, then we have in a definite sense confirmed the existence of the underlying realities, relationships, and processes we have hypothesised—or of something very much like them.

5. The laws of nature as they actually function

As we have already seen, the laws of nature as they actually function take us beyond 'our laws of nature'. They consist of the physics, chemistry and biology we more or less understand, the physics, chemistry, biology, psychology and sociology we do not understand, and the other regularities and relationships which fall outside the competencies of the natural sciences.

These would include those which involve the personal, the metaphysical, the theological (perceived revelatory), the special and the particular. These last are referred to by Albert Borgmann as 'the deictic'—those aspects of reality which can be pointed out or indicated but not understood in terms of general patterns or laws.[5]

Among this class of regularities and relationships which fall outside the perview of the natural sciences are such important human or spiritual realities as values, and the absolute claims of truth, goodness, beauty and love. They are essential for living our lives meangingfully as human beings, and for culture and society. Where do they originate? What is their basis? It can only be certain deep relationships with transcendental realities we dimly intuit.

This leads us to recognise that there are indeed important aspects of reality which are hidden or at least partially veiled.

    Dame Press, 1988), 1-47; *The Inference That Makes Science* (Milwaukee: Marquette University Press, 1992), 112 .

5.   Albert Borgmann, *Technology and the Character of Contemporary Life* (Chicago, IL: University of Chicago Press, 1984).

William R Stoeger SJ

Even such apparently scientifically accessible relationships as the dependence of life and consciousness on the complex organisation of material reality—despite its transience and fragility—seem impossible to adequately model. And when we try to understand the meaning of suffering and death, or the pervasive transcendent human experience that life, consciousness and communion are dominant and have priority, despite transience and death, we realise even more that there are key relationships and aspects which are beyond what we can comprehend. Thus, there are indeed regularities, processes, and relationships—laws of nature—which are not only beyond the competency of the natural sciences, but which are, at least for the present, beyond our knowledge and understanding.

Thus, when we speak of certain real or conceived events as being 'violations of the laws of nature', we are really only considering them as such with respect to 'our laws of nature', those we have modelled and understood to some degree. However, it is likely that such events, whether they be personal, metaphysical, or theological, would not be 'violations' if we understood all the laws of nature as they actually function, particularly the way certain 'higher' regularities and relationships having to do with the personal, the social, the transcendent may in some cases override, modify, or subsume the laws of nature we express and model in physics, chemistry and biology. It is very clear that, in light of contemporary natural sciences and our philosophical and theological reflections upon them, God acts primarily—if not exclusively—through the laws of nature as they actually function, including those which operate in evolutionary processes.

But, obviously, there are many levels of these laws of nature we do not adequately understand. What we may be tempted to interpret as divine intervention, such as cases of God's special divine action, may in fact simply be God acting through patterns, regularities and relationships—including God's personal relationship with those who are open to God's presence and initiative—which are beyond 'our laws of nature'.

6. Divine action

In light of our discussion of the laws of nature, what can we say about divine action? First of all, it cannot be like the action of any other causal factor. Secondly, it endows everything else with existence and order and with the capacity for relationships. Thirdly, it is richly differentiated, that is, God expresses God's self, and enters into or invites relationship, differently with respect to each created entity or system. Fourthly, God's action is both radically transcendent and immanent with respect to what is not God. The transcendent character of God's presence and action means not only that it is unlike any other cause (the first characteristic above), but also that it knows no barrier or obstacle and cannot be objectified. It is beyond anything we can grasp. But the immanence of God's action signifies that it is present and active—in highly differentiated ways—at the very core of everything that is. It is God's transcendence which enables God's active immanence. Finally, God's action is radically trinitarian—it is effected and realised through the dynamic relationships of Father, Son and Holy Spirit with one another (*ad intra*) and with created things (*ad extra*).

God, therefore, is present and active in all the regularities, processes, relationships and structures of reality—those we have adequately modelled and those we have not. The 'channels' of God's action in the world and the universe are always the laws of nature as they actually function.

God's action is not restricted by the constraints imposed in our inadequate and provisional descriptions or versions of them. God acts and continues to act through *all* the relationships and processes which exist, including the personal, the metaphysical, the 'deictic'—and even through chance, randomness, tragedy and accident.

William R Stoeger SJ

Finally, there is the issue of direct and indirect divine action. By 'direct divine action' I mean God's action without mediation by some other agent or cause—without some channel. In most cases, it seems that God acts through some other agent, or cause—secondary causes—through a prophet, an event, a created manifestation or symbol. Otherwise, we would not be able to notice or be influenced by the divine action. However, even with such indirect divine action, there seems to always be a step which involves direct divine action. How does God directly influence, inspire or trigger secondary causes? At some stage it seems that such direct communication from God is necessary. How is this direct link realised? This is the issue of 'the causal joint'.

My suggestion is that this link is precisely God's transcendently immanent, highly differentiated presence in each secondary cause—at the core of the being of the agent or cause—not simply holding it in existence but being present to it, and inviting its response, in the way and at the level corresponding to its complexity and receptivity.

If we understand divine action within this overall framework, then we can begin to see how God's special action in history can probably be understood as being intimately connected with God's universal creative action. God's special action in the events and relationships of history involves freely acting human beings—individually and corporately—who are either sensitive or insensitive to the initiatives and invitations God has made to them, directly and through their relationships with one another and with the world around them. The results of those human responses and what God is able to effect through them are, at the level of conscious, free human response, the extension of God's immanent creative action in the human, social-political-economic sphere. It is through this level of divine action that God is, with creation's cooperation, struggling to bring all things to completion in the Word and the Spirit. This completion will be realised in the eschatological fulfilment of 'the new heavens and the new earth', which will

Science, the Laws of Nature and Divine Action

be the fullest expression of God's creative action—of the relationship of God and creation—of all created beings as individuals and as a community—with one another in perfect communion. As John Polkinghorne has emphasised, this fulfilment will be in continuity with the creation we experience now in many ways—it will be the same creation and the same individuals who belong to it.[6] But there will be significant discontinuities from what we experience now. The physics, chemistry and biology of the new heavens and the new earth will not be quite the same. The laws of nature will be significantly different, since transience and death will have been conquered by God's relationship with us and with the rest of creation. Creation as we experience now is incomplete—its eventual completion will be effected by God through laws of nature as they actually function—and by God operating in and through them.

---

6. John Polkinghorne, 'Eschatology: Some Questions and Some Insights from Science', in *The End of the World and the Ends of God*, John Polkinghorne and Michael Welker, editors (Harrisburg, PA: Trinity Press International, 2000), 29-41.

Cosmology and a Theology of Creation

William R Stoeger SJ

What do the sciences find when they study the universe? And *what* is the universe that cosmology studies? The answers to these two questions provide us with some of the data upon which we must reflect theologically, along with the data of revelation itself, to construct a more adequate theology of creation. Up until very recently this area of theology has been sorely neglected—creation, by and large, had been considered secondary, a mere stage upon which the important drama of redemption is enacted. But now we are beginning to realise that both creation and redemption are important, two intimately connected moments in God's revelation of God's self, which continue throughout time. It is often pointed out that we come to know God and the truth of reality through the two books—the book of nature and the book of the revealed word of God, the Scriptures—and that these two books complement one another. We now look more closely at the book of nature as we read it with the help of the natural sciences, to see what it can tell us about creation and its relationship with God.

1. What do cosmology and the other sciences tell us?
When we look at reality through the eyes of the sciences, what do we see? We behold a extremely vast, intricately ordered universe, which consists of many different types of objects constituted through the underlying, often very complex relationships of their component objects with one another. And there are many levels of such constitutive relationships, from the component quarks making up the protons and neutrons and mesons at the level of fundamental particles, to the constitution

Cosmology and a Theology of Creation

of complex molecules from the many different atoms of the nintey-two natural chemical elements, to the cell, which is an extremely complex entity requiring the coordinated interaction of many different types of sub-systems through their molecular (enzyme and other protein) functional links. At the macroscopic level we have the full range of living creatures, trees, plants, insects, birds, animals, human beings, as well as the artefacts they make, and the various rocks, landforms, mountains, seas, lakes, rivers, beaches, and the larger islands and continents of our earth. Astronomically, we have our sun, with the earth and the other planets of our solar system orbiting it, and the moons orbiting the planets. Scattered in orbit around the sun are also innumberable asteroids, comets, and rock fragments and particles of dust.

But our sun is only one of over one hundred billion stars in our galaxy, the Milky Way. Many of these stars are grouped in open and globular clusters, some of which may contain more than 100,000 stars. And our galaxy is part of a group of galaxies, the Local Group, which consists of Andromeda, the Large and Small Magellanic Clouds, and dozens of dwarf galaxies. There are much larger clusters of galaxies—some called Abell clusters have as many as several thousand galaxies all gravitationally bound to one another—and all whirling around their common center of mass. These clusters of galaxies are, in turn, grouped into superclusters, long filamentary structures extending tens of millions of light years through space and encompassing voids, regions in which very few galaxies are found. A typical void would be of the order of one hundred million light years in diameter. Thus, everything we see, including ourselves, is made up of other things linked to one another in very special ways, according to the laws of nature. And those components, likewise, are constituted by subcomponents which are specially organised in a different way—and so on. This is often referred to as 'the hierarchical structuring of reality'. If we want to avoid the 'hierarchical' image, we can say that the objects that make up reality are all constituted by nested clusters of components, each level of clustering being differentiated from the others by specific types of entities (fundamental particles, atoms, molecules, cells, stars, galaxies) and by specific types of relationships among the components (gravity, electromagnetism, various types of genetic, biological or neurological connections and intercommunication).

The universe as we know it, and everything within it, is also constantly changing and evolving. The universe was not always the way it is now. There was a time when there were no human beings on earth, and an earlier time when there were no primates. At even earlier times, there were no animals, just primitive forms of life. And there was a much earlier time, perhaps the first eight hundred million years of the earth's 4.8 billion year history, when there was no life at all. Consider the universe itself, which we now know is between twelve and sixteen billion years old, since the big bang. We know that there was a time, the first several tens of million years, when there were no stars or galaxies, when the universe was just an expanding ball of gas, gradually cooling and only very gradually condensing into clusters of lumps. We have detected and studied with great precision the radiation from that primordial ionised gas—it is called the cosmic background radiation, 'the afterglow of the big bang'.

This cosmic evolution was an absolutely essential prelude to biological evolution—on this earth and perhaps in many other special locations in the universe. Without the formation of stars and galaxies, we simply would not be here. It is only in stars and through stellar processes that the elements heavier than helium and lithium were manufactured. And it is only with those heavier elements, such as carbon, oxygen, copper, iron, phosphorus, and so on that interesting and complex chemistry—and biology—became possible. When the first generations of stars died and spread their newly manufactured heavy elements throughout space by supernova explosions and stellar winds, it became possible for the first time for dust particles, meteorites, comets and planets to form in the vicinity

Cosmology and a Theology of Creation

of the new generations of stars. For the first time it also became possible to form complex molecules with their tremendous variety of characteristics. Thus, in the cooler, chemically rich environments of rocky planets and comets around stable, low mass, long-lived stars, chemical, biochemical and biological evolution became possible. That is what happened on our planet, the earth.

As the primordial, nearly homogeneous gas throughout the infant universe expanded and cooled—before there were any stars or galaxies—it eventually cooled below four thousand degrees Kelvin, and became neutral—neutral hydrogen and helium. Above that temperature the primordial gas was ionised, a mixture of free electrons and protons. The free electrons interact very strongly with photons—the particles that constitute light—causing the matter in the universe to be strongly coupled to, and opaque to, the radiation. Once the temperature of the matter in the universe fell below that four thousand degree transition region, the matter decoupled from, and became transparent to, the radiation. From that time on, the matter—in particular the matter made of protons and neutrons (baryons)—was free to clump and form galaxies and stars and star clusters. But this clumping only happened gradually. The slightly denser gas clouds expanded more slowly than the background space—because the larger density induced a larger gravitational force which pulled the clouds together more strongly and resisted the expansion. Thus, these primordial gas clouds gradually increased in density, and their rate of expansion slowed even further. Eventually, they stopped expanding altogether, even though the universe kept on expanding, and began to collapse under gravitational attraction, heating up and fragmenting to form clusters of stars, galaxies and clusters of galaxies.

In order to properly account for cosmic, physical and chemical reality as we have come to know it through observation and theory—in order to render it intelligible—we have had to invoke other processes very early in the universe.

In particular, it is clear that the universe was once very, very hot and very, very dense. In fact, the earlier we observe the universe, the hotter and denser it is—we must remember that, as we look out into the universe farther and farther, the farther back in time we see it, and the hotter it is. This is because the speed of light is finite, and because the universe is so extensive that events which occurred in a very distant star or galaxy billions of years ago are only now being seen by us. It has taken all that time for those signals to reach us. The past limit of these hotter, denser phases as we go backwards into the past we refer to as the big bang, or 'the initial singularity'. In the extremely hotter, denser states right after the big bang, mass-energy behaved quite differently than it does in the cooler more familiar regimes of today. There was a time when it was so hot that there were no protons or neutrons, just a quark sea mixed with things like electrons. Earlier than that things were so hot that there was not a difference between the two nuclear forces and the electro-magnetic force—this was the region of 'grand unificiation'. Even earlier, in what is referred to as the Planck era, it was so hot that there was no distinction between gravity and the other forces—this is the realm of total, or super-, unification. Space and time as we know them did not exist in the Planck era—everything was in an extreme quantum state. Shortly after this Planck era, it is very likely that there was a very brief but very significant period of 'inflation', a period of extremely rapid expansion and cooling, during which the size of the universe increased by as many as seventy orders of magnitude. That was followed by a period when it exited its inflationary state—with a markedly reduced rate of expansion and reheating. All these processes—or processes much like these—must have occurred to explain the universe as it is now.

Thus, there are levels of intelligibility which characterise reality and our understanding of it. We need the big bang, unification, and inflation to make the vast, expanding and cooling, causally connected observable universe intelligible and understandable—for it to make sense. And we need the

Cosmology and a Theology of Creation

primordial gas clouds to make clusters of galaxies and galaxies intelligible, and, in turn, we need those clusters of galaxies and galaxies to make clusters of stars and stars themselves intelligible, to make the solar system intelligible. And we need the solar system—and stellar systems in general—to make the earth and the planets, and all the things that develop on them, intelligible.

Looking at the observable universe, too—everything that we can see, or will be able to see, to the limit of our probing—we begin to realise that it is only intelligible if it is part of some much larger 'universe as a whole'. It is this universe as a whole—not just the observable universe—which cosmology studies—that is, it is the complexus of all those structures, conditions and processes which explain the observable universe. This must be more than, and greater than, the observable universe itself. That universe as a whole, from a philosophical pespective, however, is only intelligible if there is some ultimate source of its being and its order. These are the levels of intelligibility which span the natural sciences and philosophy.

Before going on to look more philosophically and theologically at the characteristics of physical reality the natural sciences disclose, we need to discuss briefly how we have come to know so much about the universe at large. The first really important observations which led to the development of modern cosmology were those of Hubble and Humason, who found that the farther away a distant galaxy is, the larger the redshift it has. The larger the redshift, the faster it is moving away from us. That is, the more distant galaxies are moving away from us more quickly. This was the first indication that the universe is expanding—and it is not simply that the galaxies are moving *within space* more quickly, but rather that space itself is expanding, carrying the galaxies with it. This observation has been repeated tens of thousands of times, on galaxies farther and farther from us. And the result is extremely well confirmed.

William R Stoeger SJ

The second observational indication that the universe is expanding and cooling, that it was much, much hotter and denser in the past, is the abundance of helium, deuterium and lithium. Let us concentrate on helium and deuterium. It is observed that about twenty-four per cent by weight of all matter in the universe is helium. It is impossible to make that much helium from all the stars that have ever existed in the universe. Where did it come from? The answer is that between about one and three minutes after the big bang, the temperature and the density of the universe were just right for making that much helium, and some deterium and lithium—at temperatures of several billion degrees. Deuterium is a very fragile isotope and tends to be destroyed by stars, not produced by them. But a small net amount, which is seen, was produced at the same time as the helium.

The third and perhaps most important set of observations concerns the cosmic microwave background radiation (CMWBR), which was first detected by Penzias and Wilson in 1965 and has been measured very, very precisely on all scales since then. The CMWBR is a bath of blackbody (equilibrium) microwave radiation which pervades the universe. We observe it coming to us from all directions and having a blackbody temperature of 2.73 K, with a variation of not more than 1 part in a 100,000 over the whole sky, once our peculiar motion with respect to it (the dipole moment) is extracted. This CMWBR originates from the hot ionised primeval gas before there were ever stars or galaxies, and was last scattered from the gas at the surface, or 'fog bank' which surrounds us, which our line of sight in microwaves encounters at the time when the temperature of the universe is about 4,000 K, about 300,000 years after the big bang. As we go back in time, that is where the gas becomes opaque, due to the ionisation. More recently very slight variations in the temperature of the CMWBR, which code for density fluctuations, have been detected and precisely measured. Those slight density enhancements, clouds in the primordial plasma, are the seeds of galaxies or clusters of

Cosmology and a Theology of Creation

galaxies. They grow in density as the universe around them expands—they expand, too, but more slowly than the surrounding universe, because of their increased self-gravity—and then eventually collapse to form clusters of stars, galaxies, clusters of galaxies, or the building blocks of these agregations.

Finally, it is worth mentioning that, with the advent of the Hubble space telescope and large ground-based telescopes, we have now detected very early galaxies, some of which are referred to as 'faint blue galaxies'. Some of these would have interacted and combined to form the types of galaxies we see much closer to us—in recent times. This is another indication that the universe has been, and is, evolving.

2. Characteristics of reality revealed by cosmology

If we distil from what cosmology and the natural sciences reveal to us about reality those essential characteristics which reflect something about how God relates or does not relate to the universe and to us, what do we find? The following key characteristcs emerge. First of all, reality as we come to know it from the sciences is radically relational—entities are related to other entities in essential and highly differentiated ways, through the laws of nature—those precisely described by physics, chemistry and biology, and those which transcend these descriptions. Secondly, a very pervasive aspect of this relationality is that, as I have emphasised above, reality is hierarchically structured—there are many nested levels of complexity, and each entity is constituted by interrelationships not only among its constituent components but also by its interrelationships with the larger wholes of which it is a part. There is generally both a bottom-up and a top-down aspect to the relational constitution of any given object. What is also remarkable is that on any level the constitutive substructure supporting that existence and those features peculiar to that lower level do not have to be taken into account in detail in order to understand or model how the entities at the higher level are related to form entities and systems at the next higher level—although they can be, if one has been able to successfully reduce them to lower level descriptions.

Thirdly, the universe—physical, chemical and biological reality—is evolving on all scales. The universe began as an undifferentiated expanding and cooling configuration of mass-energy. As it expanded and cooled, new things occurred—it complexified microscopically and macroscopically, fragmenting and clumping into stars, clusters of stars, galaxies and clusters of galaxies, and within and around these generating all the elements, which in some venues combined into complex molecules, and even more complex living organisms. And, once life emerged on Earth, that life evolved remarkably into a marvellous succession of genera and species, within their ecological and social systems.

A fourth and extremely important feature of reality is that it is formationally and functionally integral and autonomous. The dynamisms within nature and within the many levels of systems constituting it—which the laws of nature describe and model—are enough to account for the evolutionary processes from which radically new entities and organism emerge. They are also enough to account for the intricate organisation and balance which characterise the functioning of the physical, chemical and biological world. There is nothing more needed from the 'outside'. Nature is complete—and needs no supplementary principles of organisation or creativity. This, of course, does not mean that it explains its own existence or order, but that, granted that existence and order, nothing else is needed.

Despite what some evolutionary biologists would say, the universe and the material world do manifest some definite directionality in their evolutionary processes. It is not a pre-programmed directionality oriented to unique and specific outcomes, but rather a more general directionality, in which chance plays a signficant role, within the larger framework of order. The overall expansion and cooling of the universe

Cosmology and a Theology of Creation

determines this general directionality towards complexification and differentiation. The universe is evolving from a hot, homogeneous (smooth), undifferentiated, and simple state into an ever colder, more lumpy, more complex and highly differentiated state. And this is occurring on all levels—not without significant temporary local reversals, however.

Sixthly, with respect to the strange and puzzling phenomenon of time, it is clear that time is very important in the universe and in reality as we know it. The universe, and the earth, were very different earlier than they are now. And it is only with lots of time that the features of the universe and of the earth we now observe emerged. And, yet, it is beginning to appear, from the preliminary findings of gravitational physics and quantum cosmology, that time is not primordially basic. Very close to the big bang, during the Planck era, it may be that time as we know it did not exist. It may have only gradually emerged from the initial quantum gravitational configuration after the big bang. It is also true, more generally, that while many aspects of the universe evolve in time, or are expressed in time, there our some—for instance the stability and coherence of the laws of nature themselves, as well as the validity of logical and mathematical principles and relationships—which are supratemporal, that is, they remain true, as far as we can tell for all times.

Closely related to the issue of time are the features of transience and finiteness. Though there are very significant, long-lasting configurations and systems in our universe, they are always changing, and nothing lasts forever. Everything has a beginning and an end. But there is a strange interdependence of beginnings and ends—it only with the demise of earlier entities (for example stars or organisms) that the emergence of new entities becomes possible. Thus, birth and death, the disappearance of the old and the surprising generation of the new, is writ large throughout creation.

Finally, it appears that nature and the universe have been fine-tuned for life. This is often referred to as 'the anthropic principle'. As we have become more knowledgeable about the detailed structure of reality and its laws of nature, we have begun to realise that changing any of them—for example altering any of the fundamental constants of nature or the parameters which describe the fundamental features of the universe or the mass-energy in it by a few percent—would make the universe very, very different—and inhospitable to life. Not only inhospitable, but completely sterile. For instance, if the initial rate of expansion of the universe had been just a little too slow or a little too fast, the universe would have either collapsed very early or expanded so rapidly that stars would not have formed. In either case, life would not have emerged. Remember, without stars, we would not have the elements heavier than helium or lithium which are essential for microscopic complexity and life. It's difficult to know what to make of the anthropic principle—it does not provide an explanation and it does not necessarily imply purpose in the universe (nor does it rule it out). It helps us understand how very specific and important are the various necessary conditions for life and consciousness, and appreciate in some detail how they have been fulfilled. It also helps us determine what other combinations of parameters (there are other ranges within parameter space, though very few) would yield a universe open to life and complexity.

3. Cosmology and theology

How can these very general and pervasive features of the universe and of nature influence theology—in particular, theology of creation? First of all, they place very severe restrictions on how we can talk about God's action in the world, including God's creative action. For example, the formational and functional integrity of nature rules out a tinkering God who is constantly intervening in nature to effect what he/she intends. God's creative action in an evolving universe demands that God's creative action be fully immanent in and respectful of the laws and processes of nature. Considering this further with respect to the thorough-going relationality and the different levels of structure evident in the natural world, we really have to say that however God immanently acts in the world as creator, that action is expressed and mediated through those relationships and levels. In fact, going deeper into the reality of God's creative action indicates that God as creator ultimately determined those relationships and levels of structure through the laws of nature—not just in a static way, but endowed them with dynamism and potentiality for development and evolution.

In many ways, these characteristics of nature—of reality—strongly reinforce central aspects of traditional theological formulations of God's creative action in the world—for example, of primary causality, creation from nothing and continuing creation. But at the same time, they require us to adopt more profound accounts of creation in terms of divine immanence, kenosis and incarnation in material reality. In short, they force us towards a more radically trinitarian theology of creation.

4. Creation

Why is there a universe as a whole? Why is there something rather than nothing? Why is there order rather than disorder? Why is the universe ordered in this particular way, rather than in some other way? What is the origin of the laws of nature? This questions cannot be answered by cosmology or the other natural sciences—and thus they reveal their limitations. And they point directly to God and God's creative action.

There are two complementary aspects to divine creation, as it is traditionally understood and as this traditional understanding is reinforced by scientific cosmology: 1) creation from absolutely nothing (*creatio ex nihilo*), and 2) continuing creation (*creatio continua*). The essential content of *creatio ex nihilo* is the absolute ultimate dependence of everything that exists on God. There is nothing that is not ultimately dependent upon God for its existence and its character. Without God, absolutely nothing would exist. In creating God does not make use of any pre-existent material which was not created by God. This ultimate and absolute dependence upon God refers to the ontological origin of everything, not necessarily to its temporal origin. There may, or may not, have been a temporal origin of the universe, or an origin of time. That is very secondary, and not essential to the affirmation of absolute dependence. Finally, in speaking of creation from absolutely nothing, we mean no space, no time, no laws of nature, no pre-existing substrate or potentiality—nothing! This is in contrast to a similar expression sometimes used by quantum cosmologists, that quantum gravitational physics can account for the creation of the universe from 'nothing'. By this 'nothing' is meant a quantum vacuum state, or the lack of a past boundary. In either case, this is not 'absolutely nothing', but rather a potentiality which is governed by quantum laws of nature.

By *creatio continua* we mean that God's creative action always continues, sustaining and conserving everything in its existence and order. God does not create in a single moment, and then withdraw. God continues to be active as creator as long as anything other than God exists. This becomes more obvious, at least in a symbolic way, with our understanding of evolution. New things are constantly emerging in nature and in the cosmos—even new types of things. They evolve from the simpler dynamic structures and relationships which existed earlier. God is acting through the laws of nature in this evolving universe to create these new things and to maintain them in existence.

5. An open metaphysics from below

In order to construct a suitable theology of creation we need a metaphysics adequate to the full range of our experience, including the sources of revelation and data originating from the natural sciences. This metaphysics must be 'open'—since our knowledge at every level is imperfect and provisional. There are always new data, new insights, new or modified conclusions being fed into our intuitions and reflections. Among the conclusions and insights we must introduce into this open metaphysics are those which we distilled from contemporary cosmology in the second section of this article.

Here I shall outline one particular proposal for an open metaphysics which may provide a more adequate support for a theology of creation. It is that given by Colin Gunton in his recent book, *The One, the Three and the Many*[1]—it is the metaphysics of 'open transcendentals'. Besides being flexible and open, Gunton's metaphysics is a metaphysics 'from below'—though it uses ideas and concepts which seem to originate 'from above', from trinitarian theological thinking. They have been chosen because they seem to reflect the basic structure of reality as we experience it. The content and nuances they possess will be continually modified in light of new experiences, new data, new insights and understandings. They are provisional, and open to reassessment and modification. These open transcendentals provide a bridge for linking the conclusions coming from the natural sciences—and from other disciplines—with the insights resulting from philosophical and theological reflection on the more transcendental aspects of our experience. In themselves they map and describe the central features and relations that constitute each entity within the larger richly differentiated unity of which it is a part. These relations are both horizontal, linking the entity with other entities on the same level, and vertical, connecting it with all other entities at levels below and above it.

What are transcendentals? They are the core features—the universal directionalities—of all beings, in terms of which we can adequately characterise them in themselves and in their fundamental, constitutive relations. Or, as Gunton describes them, 'they are the true marks of being, within whose

---

1. Cambridge: Cambridge University Press, 1993.

conceptual dynamic we may conceive both time and eternity in their relatedness'.[2]

The specific open transcendentals upon which Gunton settles are: *perichoresis, substantiality* and *relationality*. Let us look briefly at each one of these in turn.

Perichoresis, a term deriving from the description of the relationship among the three persons of the Trinity, signifies the ontological interdependence, reciprocity and interanimation of individual entities, through their constitutive relationships. Each thing is constituted and maintained in its particularity by its relationships to other entities—those which are part of it as well as those whose influence are essential to its existence and function.

The second of Gunton's transcendentals is *substantiality*, which describes the particularity and distinctiveness of each entity. It is derived from the idea of the Spirit, which grants each being its distinctiveness and relative autonomy. This obviously complements *perichoresis*. Though any given object, organism or being depends radically on others for its existence and characteristics, it possesses an individuality, a distinctiveness and a certain freedom to be and act in its own way.

Relationality is the third open transcendental. It signifies that all being is oriented not just towards itself but to others, as well as to the whole. Communion is central to reality as we know it. *Relationality* differs from *perichoresis* in affirming that relationships are not just for the mutual constitution of entities but for establishing communion and an orientation towards the whole. The importance of ecology, symbiosis, coevolution and community are reflected in this transcendental concept. The other is central not just to our being, but to the signficance of the whole.

---

2. *The One, the Three and the Many* (Cambridge: Cambridge University Press, 1993), 157.

Cosmology and a Theology of Creation

6. Theology of creation

What shape does theology of creation take in light of what we have concluded from cosmology and the other natural sciences filtered through this open metaphysics of the transcendentals of *perichoresis, substantiality* and *relationality*? Relying also on what we know about God as Trinity from revelation, we detect a deep coherence between what is suggested there and the key characteristics of reality which have emerged from our reflections on cosmology. Obviously *relationality* as we observe it on so many different levels is central, and is somehow intimately connected with the *relationality* revealed as central to God's being. There are also strong indications that the relative autonomy and the formational and functional integrity of nature correspond to and support the radical substantiality of each entity and must frame our understanding of the meaning of the transcendence and immanence of God. With respect to the temporal and the supratemporal hinted at in nature, and certainly considering the evolving character of all that is, there is certain harmony of this with the fact that, in some way, God is beyond time. And yet God's own constitutive relationships are expressed in finite matter and in time in ever fuller and more complete ways, in emerging creation, in life and consciousness—in the incarnation of the Son of God in Christ, in the Spirit animating the world and then drawing it back to God as Trinity in complete communion.

Obviously, there is much yet to be done. It is not enough to sense or indicate these harmonies and coherences. That is just the beginning. We must show how the relationalities involved are connected—how the divine relationalities, though radically different from those in nature, also radically support and induce those which are created and how those which are created enable the revelation and expression of the divine relationalities. We must construct similar adequate links with regard to substantiality—the integrity of both God and created entities and systems—and with respect to the connections between the temporal, the dynamic and the evolving and the eternal, the divine and the accomplished.

We can make a few stuttering assertions which may move us towards a more adequate theology of creation, in absence of the adequate bridgework that remains to be done.

1. God's priorities are giving and sharing, and relationality (love and communion), both inside the Trinity and outside the Trinity. We know this from revelation, but we also see it strongly reflected it the structure of nature and the cosmos.

2. Creation is 'one act' from God's perspective, but it is spread out in time from our point of view. It is a radically interrelational act, both at God's level and at ours.

3. Created being and time, and relationships, are *finite expressions* of God's being, time and relationships which are constituted 'independent' of God—that is, not God and each instance possessing its own nature or substantiality. But they are destined, with our free response in love, to be taken up into God's trinitarian relationships and 'time'.

4. The Spirit of God is sent forth to enable the Word to be spoken and realised in richly diverse ways—each with its own autonomy and integrity, and dynamism, but also with its own network of relations with other created beings and with God.

5. Creation responds continually to God in many different ways, depending upon the capacity and complexity of each being.

6. God through the Spirit and the Word enters progressively more deeply and fully into creation (incarnation and all that flows from it), and invites fuller and fuller communion and life (resurrection, consumation, completion).

7. All creation until now is groaning in one great act of giving birth—the life, suffering and death of Christ, the life, suffering and death of humans, and the life, suffering and death of all creation is basically the same mystery connecting God and creation. 'Commitment' is ultimately expressed in death (surrender) leading to new life and fuller communion.

Cosmology and a Theology of Creation

We might summarise all this as follows: God is not above and beyond us, but right in the midst of the universe, our world and our being, as Creator and Lord. God is present, active and struggling in the heart of all that is—in evolving processes, regularities, relationships, to create, fashion, complete—all the while reverencing and rejoicing in the autonomy, beauty and emerging possibilities and novelty of creation. The big bang, the formation of the first stars, and even of our galaxy, our sun and our earth, may seem of remote interest to us today. But without them, and without many other events, processes and transitions, we simply would not be here!

It is certainly true that all that exists depends utterly upon God. But God has endowed nature with potentialities, dynamisms and relationships which develop and evolve. And thus we also depend utterly and completely on the cosmic, terrestrial and local environments which envelop us and nourish us. In fact, God continues to create and draw all things to completion through these intrinsic dynamisms, processes and relationships, which God has established with us and for us through the Word and in the Spirit.

Response to William R Stoeger SJ

Denis Edwards

I would like to make some comments on three parts of Bill's paper: 1. his description of the *general picture* of the universe, 2. his understanding of the *characteristics* of the universe and 3. the *theological connections* he makes.

1. The worldview of the expanding universe
Bill puts before us a picture of an observable universe that contains something like a hundred billion galaxies. Each galaxy is made up of enormous number of stars. Our own Milky Way Galaxy has something like a hundred billion stars.

He tells us that the universe began from a very simple and very hot state twelve to fifteen billion years ago and that it has been expanding and cooling ever since. As the universe cooled new things became possible—galaxies and stars. In the stars new elements were formed from hydrogen, including carbon and the other elements necessary for life. In at least one planet of one star, life emerged about four billion years ago. It evolved into intelligent, culture-bearing creatures capable of observing the universe that surrounds them to a distance of twelve billion light years, and capable of raising questions of meaning.

This is an astounding story. The size of the observable universe is mind-numbing. Its size and energy can appear alien and hostile. Yet Bill also tells us that it is only a universe of this kind, with this size and age, that could produce creatures like us. We depend upon the whole process—the galaxies, the stars, the elements produced in stars, and the whole long process of evolution of life on earth.

Response to William R Stoeger SJ

At the end of his life, the great theologian Karl Rahner reflected on the way that the size of the universe and the story of evolution can make us feel 'lost in the cosmos'. It produces disorientation and a 'cosmic dizziness'. Rahner had long built his approach to theology on the conviction that God is unfathomable mystery.[1] He always pointed out that in theology we must speak of God, but we stammer with words about that which cannot be expressed in words. We offer our thoughts on the God who is utterly incomprehensible. Rahner suggests that the experience of the scientific story of the universe can lead to the good result of a new and more profound sense of our contingency and finitude. It can lead to a deeper sense of our creaturehood and of God's mystery. Rahner considers that 'the feeling of cosmic dizziness can be understood as an element in the development of people's theological consciousness'.[2] It can lead us to stand like Job before a creation that staggers us and before a Creator who is incomprensible mystery.

Bill has deepened this sense of mystery when he shared with us the working assumption of many cosmologists that the universe we observe is but a small part of a much larger universe as a whole. They assume that the universe may in some sense have always existed. It may be infinite in extent. It may involve many universes. These ideas, of course, extend the universe as a whole beyond anything we can begin to comprehend. Theologically it takes us back to Job with an even more radical sense of the dark mystery of the creator, who creates such a universe, and who according to the Christian conviction approaches us in love and cares tenderly for each flower, bird and person on earth.

---

1. See, for example, Rahner's foundational article, 'The Concept of Mystery in Catholic Theology', in *Theological Investigations*, volume IV (New York: Seabury Press, 1974), 3-73.
2. Karl Rahner, 'Natural Science and Reasonable Faith', in *Theological Investigations*, XXI (New York: Crossroad, 1988), 50.

Denis Edwards

2. Characteristics of the universe

One of the characteristics of the universe that Bill mentions is the fact that it is relational and hierarchically structured. This means that every entity or process is intrinsically related to every other entity or process—in differentiated ways. Each thing is constituted by more fundamental entities organised through complex relationships. And each is connected to others to constitute a larger system or more complex entity. Bill describes this as operating at all levels—quarks, molecules, organisms, families, ecological systems and so on. I find this picture important in shaping the way I see reality. It is profoundly congruent with a trinitarian view of God, in which God is understood not as an individual substance, not as some great individual, but as Communion. It is also congruent with a view of human beings which sees the human not as an isolated self-conscious individual but as one who is essentially related to others, to other humans, to other creatures on earth, to the unfolding universe, and to the divine Communion that makes things be.

A second characteristic is the functional integrity and the self-determination that characterises the emergence of the universe. The universe can unfold to some extent on its own. It has a certain autonomy. It does not require outside intervention. I believe that this too has important connections with theology. It suggests a God who upholds and enables the whole process from within, but in a way that respects the proper autonomy of created processes. It suggests a God who makes room for the otherness of creation and is freely self-limiting in allowing creation its own chance to be and become.

A third characteristic of the universe that Bill mentioned is directionality. While he holds that science does not give secure support to the argument for design, he believes that there is a case for directionality in the evolution of the universe at large. He points to the anthropic principle—the fine-tuning of the universe that is necessary if humans are to emerge in it. With Bill, I do not believe that the anthropic principle can be used to

Response to William R Stoeger SJ

prove that the universe is designed by the Creator for human beings. But what it does do is to help us understand ourselves modestly but realistically as intimately interconnected with the fifteen-billion-year-old universe. We are children of the universe, individuals in whom the universe has come to self-consciousness.

The final characteristic is that of time. Bill tells us that time was not a basic characteristic of the quantum configuration that gave rise to the big bang. I find this extremely interesting and thought-provoking. He points to the fundamental importance of time in an evolving universe. Everything takes time. Everything depends on time. This makes me think about the biblical view of God as the God of promise, Jesus' reign of God, the new creation. The biblical and Christian view of God is what Karl Rahner called the Absoltute Future. The promise is that participating in the divine Communion is the Absolute Future of the creatures that evolve in time.

2.1 Reflections on the Creator
Bill tells us that the consonance we seek between a scientific view of reality and a theology of creation can find full and adequate expression only in a radically trinitarian approach. I can only completely agree with this assessment and with Bill's idea that this God works immanently and respectfully in all the processes of nature.

Bill goes further and takes up suggestions from the theologian Colin Gunton for core features of reality, which Gunton calls open transcendentals. He suggests three of these, perichoresis, substantiality and relationality. I find perichoresis and relationality impossible to separate, and would prefer to collapse these together. And I would like to add another—that of becoming.

2.1.1 An inter-related universe
If God exists only as communion, as Persons-In-Communion, then this suggests that inter-relatedness is not simply a quality

of particular beings, but is of the very essence of things. It suggests that to be is to be in relationship. This supports a view of reality as radically relational. It supports the idea that the very being of things is relational, springing from a God who is Communion. In this view, the creator Spirit of God can be understood as the Bringer of Communion, who brings forth from within the divine communion an interconnected and interdependent universe of creatures.

2.1.2 A world of diverse and distinct creatures each with its own integrity
A theology of God suggests that diversity in community may be the very nature of all reality, uncreated and created. The divine Communion does not mean the obliterating of differences but the flourishing of diversity. This suggests that creation might best be understood as a world in which each diverse creature has its own distinct integrity.

This seems congruent with a scientific view of reality in which individual creatures are radically interconnected with others, yet have their own identity and unique autonomy. Individual entities have their own degree of self-directedness, whether we think of human beings with their experience of being free agents, of birds with their glorious freedom in flight, or of particles like photons whose individual motion cannot be predetermined.

3. A world in the process of becoming
Individual creatures exist only in time between the past and an unknown future. The atoms that make up the bodies of birds, fish and mammals have been produced in a process that had to take many billion years. Evolution on earth has taken three to four billion years. The universe evolves over extraordinary lengths of time, and without the patiently unfolding of things in time, with all the regularities of nature and a great many contingent events, nothing could happen. Individual entities exist not only in communities of inter-relationship in the

Response to William R Stoeger SJ

present, but they have a relationship in time with all the creatures that preceded them, and with all the unknown creatures that will follow.

Karl Rahner was right to insist that the discovery that we are part of an evolving world demands a new understanding of reality, a new metaphysics. God, now, must be understood not simply as the dynamic cause of the *existence* of creatures, but as the dynamic ground of their *becoming*. In all events the Spirit acts to bring creation into a new future. The Spirit enables the new to occur in time. The creator Spirit is immanent in a time-bound universe, deeply involved with its becoming. But the Spirit is also always transcending of time. The Spirit is the eschatological Spirit, the Spirit of the divine eternal Communion. The biblical sources of Christian theology teach us to value and love the present moment as the gift of grace. They teach us to look back on the past and to hold it in memory as a story of what God had done for us. But in a unique way the Scriptures orient us on God's future for us. For Israel this took the shape of the divine promise, prophetic hope, and messianic expectation. In Jesus it found expression in his preaching and praxis in the light of the coming reign of God. For those who follow the way of Jesus, and live in the light of the promise of resurrection and new creation, it means living in constant expectation of God as the Absolute Future. In a recent theology of evolution, John Haught has argued that we can no longer afford to operate with a metaphysics of the 'eternal present'. He seeks to contribute to a 'metaphysics of the future', arguing that we need a view of reality that makes sense of the fact that evolution brings about *new* forms of being. He sees God as *the power of the future*.[3] The entire universe is always being drawn

---

3. John F Haught, *God After Darwin: A Theology of Evolution* (Boulder, Colorado: Westview Press, 2000), 81-104. In building a concept of God as *Absolute Future*, Haught engages with the work of Teilhard de Chardin, Karl Rahner, Jurgen Moltmann, Wolfhart Pannenberg and Ted Peters.

forward by 'the power of a divinely renewing future'. The Spirit of God is the Power of the Future, immanent all of the processes of the evolving universe, enabling it to become what is new.

God, Process and Cosmos: Is God Simply Going Along for the Ride?

Mark Wm Worthing

One of the most difficult and frequently encountered problems in the science/theology dialogue is that of the nature of the ongoing relationship between God and the physical universe. How do we understand God's relationship to an evolving universe? Does or can God act in any real sense in the processes of the physical universe? Or, to put the question bluntly, is God just going along for the ride? On the one hand, we are faced with the problem of a God constantly intervening or tinkering with the creation, after the manner of Newton's cosmic plumber. What kind of God would this be? And how could legitimate science ever take place under the shadow of an omnipotent being constantly changing or breaking the rules? On the other hand, if God does not or cannot intervene at all—say for instance to take on human form or to raise someone from the dead—then where does that leave the fundamental truth claims of Christian theism? Either way we face a very real dilemma. If we admit to intervention then it appears that we undermine science. If we deny intervention, then it appears that we erode the very core of Christian faith.

At this juncture it may be tempting to suggest that the idea of a transcendent creator capable of intervening in the physical universe is outdated. We may further be tempted to conclude that traditional affirmations about God such as omnipotence, transcendence, and the related doctrine of creation *ex nihilo* must be abandoned in favour of a more compassionate and vulnerable God who is in process along with the physical universe but cannot control its outcome. Such a view of God, we might argue, leaves room for modern science while not denying Christian faith. Yet in such an approach would too much be striped away from God in order to make way for scientific enquiry?

While the problem of divine action and physical law will always elude easy answers, the problem can at least begin to be addressed by clarification of terms. If we suggest that at heart the difficulty lies with the apparent incongruity of divine intervention and natural law then we must begin by clarifying what we mean by both concepts.

The language of divine intervention within the context of a scientific world-view is inherently problematic. In fact, the term 'intervention' is loaded with so much baggage it is best avoided in most cases. Yet I have chosen deliberately to make use of it in this discussion to heighten the extent of the difficulty and to avoid solutions that appear to reject the possibility of a God who acts. The dictionary definition of intervention is that act of entering or occurring extraneously. Immediately we have a difficulty. Can God ever act or be 'extraneous' to God's creation? If God is integral to the processes of the physical universe, then can divine action ever really be true intervention—as if by a third or unrelated party?

In theological discussions about divine intervention one usually gets the sense of at least two different uses of the word. In the stronger or more radical sense an act of divine intervention is one in which God interrupts, sets aside or violates natural laws or processes in order to bring about a divinely willed outcome. If we insist on this sense of divine intervention there would appear little room for dialogue. There is, however, a more integrated use of the concept that views divine intervention as God working with and through natural processes and laws—in ways we cannot fully comprehend—in order to bring about an outcome that would not have occurred without some 'action' on God's part. In this latter view intervention need not and must not be seen in opposition to the laws of nature that God established when God called the material universe into being. Yet at the same time, divine action is more than God hoping for the best, or simply going along for the ride. God remains God in relation to the creation that God calls into being and sustains. But given the tendency towards understanding intervention in the stronger sense as a violation or interruption of physical laws, it would seem that at the very least the term should be used cautiously.

But, of course, tinkering with the definition of 'intervention' and seeking more subtle understandings of what this may or may not entail does not entirely overcome the problem. We are still left with the difficulty of what we understand by the concept of natural laws. These, too, taken in a rigid sense, can appear to leave little or no room for divine action of any description. But if the concept of divine intervention has been prone to misunderstanding, the concept of natural law has suffered no less from inflexible representations of what it may or may not allow.

A century ago the scientific community seemed agreed that the universe was static, without beginning or end, and governed by a set of unchanging, unbending physical laws. In such a context any sort of divine action smacked of the most crass variety of interventionism. Just what role a transcendent creator could play in such a universe remained to be seen. Yet we now view the universe much more fluidly. Like the life it contains, the universe, too, is in process, engaged in a dynamic dance of cosmic evolution. In the context of the current view of the universe, the question of divine action must also be a much more fluid one. A survey of the current situation in the natural sciences suggests that the apparent necessity to exclude all forms of intervention may not be so necessary as it first appears. Three areas of contemporary science merit special attention in this regard. These are the uncertainty principle of quantum mechanics, current understandings of the nature of physical law, and chaos theory.

But a word of caution must be spoken before we look at these three areas. Our intent is to demonstrate that it may not be so easy to say in all cases what may or may not be allowed by natural law. The laws of nature, like the universe they govern, are more complex than they once appeared. Yet pointing to this complexity must not be seen as making room for God to work. The assumption that God can only be seen to act when and where natural law is unclear or imperfectly understood would be a rather unfortunate return to a god-of-the-gaps mode of thinking. God is able to act not simply because physical law is more complex and more 'flexible' than we are sometimes prone to think. Rather, a more realistic assessment of natural law—like a more realistic assessment of divine action—demonstrates the difficulty of assuming that God's action in the physical cosmos is simply not possible.

1. Quantum mechanics and the uncertainty principle

An explanation of precisely what quantum mechanics is cannot be attempted here, nor is it necessary for our purposes. Instead, I would like to highlight one of the more famous insights to emerge from the development of this field of science, namely, the uncertainty principle.

Laplace, the French scientist famous for allegedly telling Napoleon, when he inquired about the place of God in his theories, that he had 'no need of that hypothesis',[1] was a strict determinist. He believed that Newton's laws enabled us in theory to predict every future event in the universe—provided we could determine with precision the complete state of the universe at any one time. Laplace believed that similar, yet to be discovered laws determined other things in the universe, including human behaviour. Needless to say, such scientific determinism was strongly resisted by theologians, who saw it as an infringement on God's freedom to act in the world. Yet scientific determinism remained the standard assumption of

---

1. Cited in Roger Hahn, 'Laplace and the Mechanistic Universe', in *God and Nature*, D Lindberg and R Numbers, editors (Berkeley: University of California Press, 1986), 256.

God, Process and Cosmos: Is God Simply Going Along for the Ride?

science up to our own era. The fate of scientific determinism was signalled when Max Planck, as a result of his explanation of the laws of black body radiation, suggested in 1900 that waves from such sources were released in packets that he called quanta. The details of Planck's theory of quanta solved the mystery of heat emission by demonstrating that a body could lose energy only in finite, extremely small amounts (quanta), and was thus widely accepted as valid. It was not, however, until 1926 and the work of another German physicist, Werner Heisenberg, that the implications of quantum theory for determinism were realised. Heisenberg, using Planck's quantum theory to determine the feasibility of measuring the velocity and position of a given particle, found that the light quantum needed to measure the particle interfered with it in unpredictable ways so that the more accurately one attempted to measure a particle's velocity, the less accurately it was possible to measure its position, and vice versa. As Stephen Hawking put it: 'Heisenberg's uncertainty principle is a fundamental, inescapable property of the world, . . . [that] signalled an end to Laplace's dream of a theory of science, a model of the universe that would be completely deterministic'.[2]

Heisenberg, along with Erwin Schrödinger and Paul Dirac, developed the theory we now know as 'quantum mechanics' based on the uncertainty principle. Their theory predicted no specific observable events, but rather a range of possible results along with formulae for predicting statistically the chances of obtaining each possible result in any given instance.

But when speaking about the grand scale of the cosmos at large and God's role in governing its processes, the difficulty of predicting the actions of subatomic particles may not seem at first particularly relevant. Yet this is far from the case. The implications of the uncertainty principle for the larger scale

---

2. For this and the above see Stephen Hawking, *A Brief History of Time: From the Big Bang to Black Holes* (New York: Bantam Books, 1988), 53ff.

structures of the universe can be best illustrated by the famous example of Schrödinger's cat.

In 1935 Erwin Schrödinger sought to demonstrate the implications of applying the uncertainty of quantum probability to large-scale structures. Schrödinger proposed a thought experiment in which the fate of a hapless cat becomes contingent upon the quantum probability of the decay of a small amount of radioactive substance that exists within a closed system that also includes the cat. Schrödinger explained the experiment as follows:

> A cat is penned up in a steel chamber, along with the following diabolical device (which must be secured against direct interference by the cat): in a Geiger counter there is a tiny bit of radioactive substance, so small, that perhaps in the course of one hour one of the atoms decays, but also, with equal probability, perhaps none; if it happens, the counter tube discharges and through a relay releases a hammer which shatters a small flask of hydrocyanic acid. If one has left this entire system to itself for an hour, one would say that the cat lives if meanwhile no atom has decayed. The first atomic decay would have poisoned it. The w-function of the entire system would express this by having in it the living and the dead cat (pardon the expression) mixed or smeared out in equal parts.[3]

Because of the superposition of various possible states in quantum mechanics, the fate of the cat, if we are to remain consistent with the Schrödinger equation, must be calculated by someone outside the closed box in the same manner as the probability of the decay of the radioactive material. In other

---

3. Erwin Schrödinger, 'The Present Situation in Quantum Mechanics', translated by J Trimmer, in *Quantum Theory and Measurement*, J Wheeler and W Zurek, editors (Princeton: Princeton University Press, 1983), 157.

God, Process and Cosmos: Is God Simply Going Along for the Ride?

words, according to quantum theory, until observed, the cat is 'in a complex linear combination of dead and alive. It could be dead plus alive. It could be dead minus alive. It could be dead plus the square root of minus one times alive'.[4] Of course, this sounds like nonsense. Schrödinger himself believed the cat would either be dead or alive and that at some point between the quantum, micro-level and large-scale reality his equation breaks down and must be adapted to complicated large-scale structures such as cats and the universe.[5] Others, however, have more confidence in Schrödinger's equation than Schrödinger himself had and believe that the cat paradox demonstrates that classical realism does not work. Paul Davies and J Brown, for instance, have written that 'the paradox of the cat demolishes any hope we may have had that the ghostliness of the quantum is somehow confined to the shadowy microworld of the atom, and that the paradoxical nature of reality in the atomic realm is irrelevant to daily life and experience . . . Following the logic of quantum theory to its ultimate conclusion, most of the physical universe seems to dissolve away into a shadowy fantasy'.[6]

The prevailing view, however, remains that of Schrödinger himself, namely that quantum physics must be adapted for large-scale systems. The interesting question theologically is whether the adaptation of the indeterminism of quantum physics to the apparent determinism of large-scale structures could serve as a model for a similar adaptation of the providential freedom of God to the apparently deterministic large-scale structure of the universe? Science, at least to the extent that it is influenced by quantum mechanics, is no longer so certain as to what can and what cannot happen.

---

4. Roger Penrose, 'Big Bangs, Black Holes and "Time's Arrow"', in *The Nature of Time*, R Flood and M Lockwood, editors (Cambridge, Mass: Basil Blackwell, 1965), 58.
5. Schrödinger, *op cit*, 157.
6. Paul Davies and JR Brown, *The Ghost in the Atom: A Discussion of the Mysteries of Quantum Physics* (Cambridge: Cambridge University Press, 1986), 30.

Mark Wm Worthing

2. Divine intervention and the nature of physical law

According to Stephen Hawking,

> science seems to have uncovered a set of laws that, within the limits set by the uncertainty principle, tell us how the universe will develop with time, if we know its state at any one time. These laws may have originally been decreed by God, but it appears that he has since left the universe to evolve according to them and does not now intervene in it.[7]

This assertion would appear to contradict the traditional Christian belief in the possibility of divine action; in particular that very special category of divine action termed miracles. For a miracle, in the mind of many, is precisely that; an interruption or exception of the physical laws that govern our universe.

The so-called 'interventionist,' special providence of God need not, however, be understood only in terms of the miraculous. It is entirely possible for a 'special' act of providence that 'intervenes' in human or natural history to take place without violating any laws of nature. Hence Arthur Peacocke is correct to contend that particular events or clusters of events, 'can be intentionally and specifically brought about by the interaction of God with the world in a top-down causative way that does not abrogate the scientifically observed relationships operating at the level of events in question'. Such a possibility, according to Peacocke, is of value in that it 'renders the concept of God's special providential action intelligible and believable within the context of the perspective of the sciences'.[8] Yet must we limit the nature of God's action in the universe to that which falls within the confines of our own

---

7. Hawking, *A Brief History of Time*, 122.
8. Arthur Peacocke, *Theology for a Scientific Age: Being and Becoming—Natural and Divine* (Oxford: Basil Blackwell, 1990), 182.

understanding of the laws of physics? Miracles, by virtue of their nature, bring the question of God's ability to act in the physical universe into sharp focus and deserve further attention.

In the strictest sense, miracles are occurrences that are not explicable within the context of presently known physical laws. It is precisely at this point that the issue has usually come to an impasse between theology and natural science. Theology has traditionally maintained that such occurrences have not only taken place in the past, but in principle, could occur in the future. Natural science has maintained that the laws of physics that govern the physical processes of our universe are invariable and, therefore, miracles are in principle impossible.

David Hume was perhaps the first, in the context of the emerging, modern scientific world-view, to deny the occurrence of miracles. Hume argued that there must be 'a uniform experience against every miraculous event, otherwise the event would not merit the appellation. And as a uniform experience amounts to a proof, there is here a direct and full proof, from the nature of the fact, against the existence of any miracle'.[9] For Hume, therefore, a miracle is excluded by its very definition. Modern science, if not individual scientists, has tended to reject miracles on this same basis.

What is at stake is not simply a dispute over individual 'miraculous' occurrences so much as the question of God's ability to act in the created order. God's general providence takes place apart from any interruption or exception of physical laws. God actively directs and sustains the universe, but within the context of the specific physical laws that God established to govern it. Yet the traditional Christian doctrine of divine providence also includes the possibility of a special providence that posits the freedom of God to act in the normal processes of

---

9. David Hume, 'Of Miracles', in *David Hume: The Philosophical Works*, Volume 4, T Green and T Grose, editors (London, 1882), 93 and n. 1.

the physical universe in such a way that these processes at least appear to be interrupted.

Perhaps the question of miracles remains difficult theologically for many because the idea of miracles has more to do with the doctrine of God and God's relationship to the physical cosmos than with particular 'supernatural' occurrences. Not only is the doctrine of miracles significant for our understanding of God, but the Christian religion is also built upon two central miracles: the incarnation of God through the virginal conception of Jesus, and the resurrection of Jesus from the dead. Clearly, Christian theology cannot reject the possibility of miracle within the context of God's special providence and remain Christian theology. But to what extent can such a special providence be maintained in the light of contemporary physics? Any discussion of miracles as the most profound form of divine action in our cosmos must collide sooner or later against the 'immutable laws of physics', which appear to disallow such occurrences in principle. It is the apparent immutability of such laws that have led Hawking and others to claim that God does not now intervene in the physical world.

Intriguingly, since the inviolable nature of physical law is presupposed by the traditional doctrine of miracles, the oft-assumed immutability of the laws of nature constitute no proof against miracles. From the perspective of theology, one might say that miracles are the exceptions that not only assume but also 'prove' the rule. Yet the difficulty is not so easily removed. Hume's criticism that miracles, by definition, cannot happen, remains a problem. Recent changes in the understanding of the nature of physical law, however, especially through quantum theory, may allow new possibilities for a theological conceptualisation of miracles in light of scientific understandings of natural law.

Given the fact that all the laws of nature have not yet been discovered or are not fully understood, there is a certain difficulty that arises in saying what they do and do not permit

God, Process and Cosmos: Is God Simply Going Along for the Ride?

with reference to the total compass of reality. The laws that describe individual systems may not be satisfactory when seeking to describe the whole.

In this light it would seem that Hawking's statement that it appears that God, if indeed a Creator-God exists, has left the universe to evolve according to the laws of nature 'and does not now intervene in it', must be seen as an observation and not made into a rule. But theology should also expect such an observation to generally hold true. After all, what kind of Creator would we confess who found it necessary to continually make adjustments and corrections to his 'good' creation. Even if a 'miracle' were verified, the 'laws' of nature could almost certainly be revised to take into account this 'new' observation as part of the 'natural' phenomena of the universe. Would it then still be considered a miracle? Do we run the risk of viewing miracles only as those things we cannot otherwise explain?

Perhaps the most radical development, however, in the understanding of the nature of physical law has been that introduced by quantum mechanics which has replaced the classical understanding of physical law with a quantum-statistical approach. Natural laws may be either universal in form and state what must happen (classical physics), or statistical in form and state what must probably happen (quantum physics). In this light a miracle, it would seem, would also be constituted by a violation of statistical probability rather than of some absolute set of laws. The precise theological and philosophical implications of such an understanding of miracles, however, remains to be seen. Yet it would seem that the recourse to physical laws as proof against any form of divine action more often than not relies upon a dated and simplistic understanding of the nature of physical law. We clearly live within a universe, the processes of which are guided by a set of physical laws (both classical and quantum statistical) that have been in place since the Planck time. What is not so clear is that divine action, including what some might call divine 'intervention' in the physical universe, must necessarily be incompatible with the existence of these laws.

3. Chaos, order and divine freedom

Into this whole mix now comes the question of order and chaos. For centuries the belief in an orderly God went hand in hand with the belief in an orderly universe. The God of Newton, whether we are consciously aware of it or not, is the God that most of us have grown up with and most easily conceive. This God is a God of order, a God who has methodically created a methodically run universe—a God who is predictable and whose world is predictable. This God, we have been taught, is opposed to the demonic influences of chaos and unpredictability.

In the last century these safe and comfortable perceptions began to come unravelled. At first almost imperceptibly. Slight tremors appeared on the landscape of physics that only a small few seemed to really understand but that have grown in magnitude to the point that we now have something akin to two world-views existing side by side in uncomfortable juxtaposition: The classical view of order and predicability and the orderly God who knowingly and sovereignly presides over it all, and the newer view of inherent unpredictability and chaos that seems to exist in paradoxical juxtaposition to the God whom most of us conceive.

The question of interest from the perspective of Christian thought is how we are to understand God's relationship to such a world. But before this question can be taken up more fully we must outline the changes in scientific thinking that have conspired to bring disorder into our smoothly run universe.

The determinism/indeterminism question, previously alluded to, is one of the big unresolved issues separating classical and quantum physics. As physicists search for a grand unified theory that would unify these two streams of physics, speculation has already begun as to the implications of such a theory for the determinism/indeterminism question. Could it

God, Process and Cosmos: Is God Simply Going Along for the Ride?

be that the unification of physics will bring with it a resolution of the tension between determinism and indeterminism? Or must either determinism or indeterminism ultimately prevail? The Oxford mathematician Roger Penrose, although himself sceptical that this will indeed prove to be the case, suggests that 'it is even possible that we may end up restoring determinism in quantum mechanics'.[10] Of course, a number of physicists believe relativity theory may well have to be adapted to make room for some degree of indeterminism.[11]

The German physical and theoretical chemist OE Rössler has written that 'it could turn out . . . that a universe that is chaotic itself ceases to be chaotic as soon as it is observed by an observer who is chaotic himself'.[12] 'Chaos', in other words, may be simply a matter of perspective. A hypothetical chaotic observer would view a chaotic universe as entirely 'in order'—and because it is observed as such it would indeed become orderly. It is an assertion of the fundamental unpredictability of the quantum, and perhaps also the large-scale systems of the universe.

The uncertainty principle of quantum physics has led, in many ways, to a re-examination of the nature of 'chaos', or what in classical terminology would be called non-linear systems. Most significant among such re-examinations have been those that propose a sort of compromise between determinism and indeterminism by suggesting that chaos is perhaps not so chaotic as we might imagine.

A great deal of attention has been given recently to the idea of a deterministic indeterminism as a possible harmonisation of the implications of classical and quantum physics. Gary Zukav, for instance, in his book, *The Dancing Wu Li Masters*, speaks of a

---

10. Penrose, *op cit*, 60.
11. Davies and Brown, *op cit*, 30.
12. OE Rössler, 'How Chaotic Is the Universe?' in *Chaos*, Arun V Holden, editor (Manchester: Manchester University Press, 1986), 317.

Mark Wm Worthing

'chaos beneath order',[13] and Ian Stewart, *Does God Play Dice? The New Mathematics of Chaos*, speaks poetically of the effort to domesticate chaos: 'Chaos gives way to order, which in turn gives rise to new forms of chaos. But on this swing of the pendulum, we seek not to destroy chaos but to tame it'.[14] On a more rigorously scientific level Heinz Georg Schuster has attempted, with his 1984 book, *Deterministic Chaos*, to analyse mathematically the phenomenon of deterministic chaos within dissipative systems. Schuster defines deterministic chaos, within the context of his study, as 'the regular or chaotic motion which is generated by nonlinear systems whose dynamical laws uniquely determine the time evolution of a state of the system from a knowledge of its previous history'. Schuster notes that due to new theoretical results and the use of high-speed computers, 'it has become clear that this phenomenon is abundant in nature and has far-reaching consequences in many branches of science'.[15]

But how would contemporary views of a deterministic indeterminism differ from a Laplacian determinism in which the future state of the universe could potentially be calculated from its present state? Roger Penrose indicates the direction in which such discussions are leading when he writes that he believes that some 'new procedure takes over at the quantum-classical borderline which interpolates between' deterministic and probabilistic 'quantum jump' parts of quantum mechanics, and that this new procedure 'would contain an essentially non-algorithmic element', which would in turn imply 'that the

---

13. Gary Zukav, *The Dancing Wu Li Masters: An Overview of the New Physics* (New York: William Morrow, 1979), 213.
14. Ian Stewart, *Does God Play Dice? The New Mathematics of Chaos* (New York: Penguin Books, 1990), 1.
15. Heinz Georg Schuster, *Deterministic Chaos: An Introduction* (Weinheim: Pysik-Verlag, 1984), 1f.

God, Process and Cosmos: Is God Simply Going Along for the Ride?

future would not be computable from the present, even though it might be determined by it'.[16]

In distinguishing between computability and determinism Penrose would seem to represent a break from a strict Laplacian determinism that did not make such a distinction. The concept of a determined but not completely predictable universe also contains possibilities for theological reflection on the providence of God. As a model of God's providential direction of the universe, Penrose's tentative outline of a 'correct quantum gravity' (CQG) theory would seem to demonstrate how it is possible to conceptualise divine providence and divine freedom as consistent aspects of God's providential sustenance of the physical cosmos. In fact, Penrose himself suggests that a CQG theory would also leave a role for human free will within an essentially deterministic universe.[17]

Similar to the tension between quantum indeterminism and classical determinism, the doctrine of God's providential direction of the universe (and of human history) has always contained a similar tension between the 'indeterminism' of human 'free will' and the 'determinism' of divine predestination. Even within the divine economy (excluding the problem of human free will) there exists a tension between the 'indeterminism' of divine freedom and divine predestination.

If God governs the creation in such a way that the universe evolves according to a strict, predetermined order, what place is left for divine freedom? On the other hand, if God is really free to act in the physical universe, what happens to the traditional doctrine of divine predestination? Theological systems have tended to ultimately opt for one or the other. In a sense, theologians proceed much like physicists in this regard. On the micro-level of miracles, prayer, human moral decision, and so

---

16. Roger Penrose, *The Emperor's New Mind: Concerning Computers, Minds, and the Laws of Physics* (Oxford: Oxford University Press, 1989), 431.
17. *Ibid*, 353ff, and 431f.

on we proceed as if God and individual persons were able to actually alter the course of history or affect in some way the state of the physical cosmos (for example, the human decision to pollute or not to pollute, to engage in nuclear war or not). On the macro-level of the teleological outcome of history or the consummation of creation the pre-determined 'plan' of God takes over. Like physicists, theologians would very much like to know how these two levels fit together. Does one ultimately prevail over the other? While current discussions among physicists cannot possibly provide answers to these theological questions, they may very well provide useful paradigms for overcoming, or at least learning to live with, the tension between indeterminism and determinism.

If physics can learn to live with so-called deterministic chaos perhaps theology can accept a providential freedom of God that accommodates the 'determinism' of divine predestination and the 'indeterminism' of human and divine freedom without denying the inevitable tension between the two or subsuming one under the other. The mathematician Ian Stewart put the matter well when he suggested: 'Perhaps God can play dice, and create a universe of complete law and order, in the same breath'.[18]

Again, the result of a closer look at the state of our current understanding of the nature of the physical universe is not that there exist holes or cracks in which some insipid god-of-the-gaps may be allowed to tinker—but rather that the nature of physical reality and the processes that govern our evolving cosmos do not appear to prohibit the concurrent free action of a transcendent creator.

4. Conclusion: Divine action as scientific and theological problem

In light of these various complexities we might ask why theists today often are loath to invoke divine intervention in the

---

18. Stewart, *op cit*, 2.

world? There is a sense in which the invocation of divine intervention has generally been viewed as a sort of theological 'cheating' similar to the invocation of a God-of-the-gaps. When all other explanations fail we invoke the miraculous intervention of God. Yet as Arthur Peacocke correctly points out, such intervention is not normally 'compatible with and coherent with other well-founded affirmations concerning the nature of God and of God's relation to the world'.[19] Hence physicist and Anglican priest William Pollard can comment that the majority of 'miracles' recorded in Scripture 'are the result of an extraordinary and extremely improbable combination of chance and accidents. They do not, on close analysis, involve . . . a violation of the laws of nature'.[20]

But for theology, it is important to distinguish between God's in-principle ability to 'intervene' in the affairs of the world through a miraculous 'interruption' of natural law and God's propensity to actually carry out such acts of special providence. From the perspective of the natural sciences it is difficult to engage in dialogue with theology if theology is constantly changing the rules by invoking miraculous intervention. Hence, partly for apologetic reasons, miracles have become something of a theological problem that contemporary theologians are 'loath to invoke'. There are also theological grounds for this reluctance. As Polkinghorne suggests, a God who is constantly tinkering with creation through special, miraculous intervention begins to look uncomfortably like a God-of-the-gaps.

Yet when all is said and done, the ability of God to act decisively in the universe remains a fundamental confession of the Christian doctrine of God. At issue is not so much the immanence of God—that can be maintained apart from God's

---

19. Peacocke, *op cit*, 183.
20. William Pollard, *Chance and Providence: God's Action in a World Governed by Scientific Law* (New York: Charles Scribner's Sons, 1958), 83.

sovereignty over the laws of nature—but the transcendence of God. A God who cannot in principle act decisively in the universe can hardly be credibly maintained to be its 'wholly other' Creator. Yet the transcendence of God is perhaps ultimately more of a stumbling block than the possibility of miracles. A God who transcends the physical universe also transcends the ability of modern science to prove or disprove God's existence. In an age in which scientific research stands on the very threshold of understanding the mysteries of the universe, a God who is beyond its grasp remains a hard pill to swallow.

Regarding the question of divine action in the light of modern science and the Christian belief in a transcendent and omnipotent God, we are left with a certain tension and uncertainty that call for restraint in our talk of miracles and other forms of divine action. To claim either too much or too little concerning the potential of divine action is to be avoided. Arthur Peacocke summarised the matter well when he said:

> Given that ultimately God is the Creator of the world . . . we cannot rule out the possibility that God might 'intervene', in the popular sense of that word, to bring about events for which there can never be a naturalistic interpretation . . . But we have . . . cogent reasons for questioning whether such direct 'intervention' is normally compatible with and coherent with other well-founded affirmations concerning the nature of God and of God's relation to the world.[21]

We must, therefore, be ever vigilant against again turning to a god-of–the-gaps. For God acts, and acts freely, not just in the seemingly inexplicable but also precisely in the more clearly understood processes of the universe. Divine action and natural

---

21. Peacocke, *op cit*, 183.

God, Process and Cosmos: Is God Simply Going Along for the Ride?

processes are integrated in ways we cannot fully comprehend. But the God who acts freely acts not just at some safe, transcendent distance, but as one who is also imminent to the creation. God is truly imminent within the cosmos but is not confined to it. Hence God also goes beyond—or transcends—that which God has called into being. And therefore a free and divine action in all aspects of the processes of the physical cosmos cannot be made to exclude what we might sometimes label as 'intervention'. A God who acts concurrently 'in, with, and under' the processes of an evolving universe is not simply along for the ride. This God *is* the ride.

Evolution and the Christian God

Denis Edwards

There is a long tradition in Christian theology that sees the capacity for scientific inquiry into creation as a God-given gift. The universe and the human capacity to explore it scientifically are understood to come from the same source as biblical revelation. For those who take this view, there is an expectation that the best work of science will reveal insight into God's creation that will be radically and profoundly congruent with Christian faith.

This does not mean that there will never be conflicts or difficulties between scientific and theological positions. But it suggests that from the perspective of this theology, there is a confidence that such difficulties will not prove terminal and that apparent conflicts are, in principle, resolvable, often in a revised theological synthesis. Sometimes further scientific work removes an obstacle, as when twentieth century science moved beyond the mechanistic model of reality that seemed to leave no room for divine action. Sometimes an advance in theology removes an obstacle, as when theology felt free to abandon the position taken by Archbishop James Usher, in his *Sacred Chronology* of 1620, that the creation of the universe occurred 4004 years before the birth of Jesus Christ.

In this spirit, I want to suggest that it is entirely possible and appropriate to embrace *both* evolutionary biology and the Christian doctrine of creation by a trinitarian God. This stance raises a number of important theological issues. Here I will consider just three of them. First, how is biological evolution congruent with the *biblical* account of creation in the book of Genesis? Second, how can we understand God working purposefully in creation in the light of the *random* nature of

Evolution and the Christian God

biological evolution? Third, how can theology think of God acting with love in the light of the *costs* of evolution?

1. Evolution and the biblical stories of creation

According to the best available science, life has evolved on earth over the last 3.8 billion years. The first living creatures were simple cells without nuclei (prokaryotes), in the forms of bacteria and archaea. These were the only forms of life for the next two billion years. About 1.9 billion years ago more complex cells with a nucleus (eukaryotes) began to appear. These gave rise to multicellular creatures, which appear in the fossil record from about five hundred and eighty million years ago. There are wonderful examples of fossils from this period in the Ediacara fauna of the Flinders Ranges north of Adelaide. There was a exuberant flowering of life in the sea during the Cambrian period (545-495 million years ago). The first vertebrates moved on to land about 375 million years ago. Life bloomed in the Triassic (245-206) and Jurassic (206-144) periods with dinosaurs, flying reptiles, marine reptiles and mammals. The dinosaurs and many other species disappeared in the extinction that occurred sixty five million years ago, but mammals flourished in the new environment. About seven million years ago, an apelike species evolved which had an upright (bipedal) style of walking, and by about 150,000 years ago the first modern humans had appeared. This is one of many lines of development from the original bacteria. Meanwhile, bacteria themselves have not only survived for the last 3.8 million years, but also been able to flourish in a wide variety of extreme environments. All creatures of earth have evolved in their own unique way from the original communities of bacteria.

How can this account be reconciled with the first biblical account of creation that tells of creation in seven days, with human beings appearing on the sixth day? I think the answer is readily available in a mainstream approach to biblical interpretation which recognises the particular type of literature,

Denis Edwards

or *literary form*, we find in the opening chapters of the Bible. I will suggest that the biblical accounts of creation belong to the literary form of primeval sagas, that such stories are not to be approached as sources for a scientific account of the history of life, but as narratives that can communicate profound religious truth concerning God as Creator.

In a modern library, books are classified according to the type of literature they represent: fiction, biography, history, poetry and so on. We expect a historical work to be concerned with a truthful interpretation of events. But the search for truth is not something limited to history. We also expect a great novelist or poet to be concerned with truth, the truth of human existence in the world in its ambiguity and beauty. We approach each different kind of literature on its own terms. Each can lead to wisdom and truth. But we would make an obvious mistake if we read Patrick White's *Voss* simply as a work of history. This might lead a reader to be misled about historical events. But more importantly, such a reader might well miss the truth of the novel. Its truth and meaning far transcend a historical work.

There is a sense in which the Bible is one book, one canonical whole. But in terms of the literature it contains, it is much more like a library. It contains diverse kinds of literature gathered from the ancient history of Israel over hundreds of years, and from the first century of the Christian church. It contains examples of epic poetry in some of the narratives of the Pentateuch, dramatic expressions of profound theology in Job, lyric poetry in the Psalms, didactic wisdom sayings in Proverbs, prophetic and apocalyptic patterns in the prophetic books, tales of tribal heroes in Judges, and the factual account of events, perhaps by an eyewitness in the court history of David (2 Sam 9 – I Kgs 2).[1]

---

1. See Raymond Brown and Sandra Schneiders, 'Hermeneutics', Raymond Brown, Joseph Fitzmyer and Roland Murphy, editors, *The New Jerome Biblical Commentary* (Englewood Cliffs, New Jersey: Prentice Hall,

Evolution and the Christian God

None of these types of literature correspond to the modern post-enlightenment view of history. It would be a mistake to read any of them as if they had the same concern for factual exactness that might be expected in a modern history. But all of them are concerned about truth, a truth that is richer and more important than factual details. I am much in agreement with the biblical scholar Roland Murphy who says that *the equation of biblical truth with historical truth is a form of reductionism.*[2] The book of Jonah, for example is a book concerned with truth, but not primarily with historical truth. If the story of Jonah is misinterpreted as history, then a reader is caught up in a debate over whether Jonah really was in the belly of a fish for seventy-two hours. But this is to completely miss the point. The book is not a historical account but a form of parable, a didactic story. This story communicates profound religious truth. Through a playful and clever narrative, the reader is led to discover that God's mercy far transcends the limits and prejudices of the all-too-human prophet. This book is a parable of the divine mercy, mercy that is radically free and unmerited. To the prophet and his people, the city of Nineveh represented all that was oppressive and hateful. That God was willing and able to show mercy to Ninevites challenges all human views of God. God's generosity is presented as beyond all sectarianism and exclusive claims. The biblical truth expressed in this story concerns the shocking nature of divine compassion that breaks through ethnic and religious barriers.

In a similar way, in order to be free to receive the biblical truth that the Genesis narratives have to offer, it is necessary to recognise that they do not have the character of modern history or science. Genesis 1–11 is not a scientific account, nor is it a modern history. Nor is it a parabolic story like Jonah. It is a primeval saga, concerned with the origins of the world, which

---

1990), 1151-52.
2. Roland E Murphy, 'Introduction to the Pentateuch', in *The New Jerome Biblical Commentary*, 5.

provides the context and foundation for the story of the origin of the Israel in chapters 11–50.[3]

The stories of origin in Genesis 1–3 share elements with neighbouring myths of origin, such as the Gilgamesh Epic, the *Enuma elish* and the Atrahasis story from Mesopotamia. The biblical stories have the same prescientific view of the visible universe as the Mesopotamian creation stories. They share a common theological concern to see the whole of life in the context of divine creation. But the Hebrew account of creation is distinctive. Here there is nothing about a plurality of gods, nothing of conflict between gods, and nothing about human beings being created to be slaves for the gods. This is an alternative theology. All of creation is now understood as springing from the one saving God of Israel. It is the liberating God of the covenant who creates in freedom and by word. The Mesopotamian myths are replaced not by a modern scientific or abstract theological account, but with another story of origin, at the centre of which is the saving, merciful God of Israel. This story communicates a profound theological vision.

It is a carefully structured narrative, which places the creation of each kind of creature within a structure of seven days, culminating in the Sabbath (1:1 – 2:3). Many biblical commentators think that this account of creation comes from the 'Priestly' source and suggest that it had its origin in or about the sixth century BCE.[4] The seven day account gives a vast

---

3. On the literary form of the Genesis accounts, see George W Coates, *Genesis with an Introduction to Narrative Literature: The Forms of the Old Testament Literature* (Grand Rapids, Michigan: William B. Eerdmans, 1983).

4. This 'Priestly' writer not only provided the creation story of Genesis 1:1 – 2:3, but also apparently incorporated and corrected an already existing creation account that had existed for several centuries. This is called the 'Yahwist' account and is found in portions of Genesis 2–8, in the stories of Eden, Cain and Abel, Lamech and the flood. The flood account reflects both Yahwist and Priestly sources. For a helpful account of all this, see Richard J Clifford and John J Collins, editors, *Creation in the*

Evolution and the Christian God

cosmic scope to God's work in creation. The whole universe and everything in it is created by God's word alone. The order of creation is as follows: the light, and with it day and night (first day); the dome of the sky separating the waters above from the waters below (second day); dry land and the sea, followed by plant life (third day); the sun, the moon and the stars, set in the dome of the sky (fourth day); the fish, the sea creatures and the birds (fifth day); the land animals and then humans (sixth day); God rests (seventh day). The creation of humans comes at the climax of this story. This account stresses the dignity of human beings, made in God's image (Gen 1:27 and 9:6).[5] All of creation is *good*. All living things receive their fertility from God and stand under God's blessing. The whole account leads to the joy, rest and peace of the Sabbath.

How is this text to be interpreted today in the light of biological evolution? I would argue that there are two ways to misread this text. A first misreading would be to presume that it can be read like a modern post-enlightenment history. The biblical authors were concerned with theological truth rather

---

Biblical Traditions (Washington, DC: The Catholic Biblical Quarterly Monograph Series, 1992). See also Claus Westermann, *Genesis: A Practical Commentary* (Grand Rapids, Michigan: William B. Eerdmans, 1987) and Bernard W Anderson, editor, *Creation in the Old Testament* (Philadelphia: Fortress Press, 1984). See also Bruce Vawter, *On Genesis: A New Reading* (Garden City, New York: Doubleday, 1977).

5. We are told that human beings are given 'dominion' over other creatures and that they are to 'subdue' the earth (1:26–30). These are harsh words, which have been misused and taken out of context to justify ruthless exploitation of the earth. In the light of the ecological crisis, it is important to recognise that this is dangerous and time-conditioned language. In its original context, it may have been an encouragement to bring human work and ingenuity to bear on the land. It is important to maintain the insight that human beings are called to cooperate with God in the unfolding of the potentialities of nature. But this needs to be understood in the context of a carefully developed ecological theology, which can build in part on texts like Genesis 2:15 and Genesis 9:12.

than with an order of events or a timetable for creation. They wanted to teach that the one saving God of Israel has created the whole visible universe. The seven day story is followed by another, the story set in the Garden of Eden, and this second story has a completely different order of events. The two stories are set side-by-side in the biblical tradition, with no hint of concern about the differences in sequence.[6] To seek to harmonise these accounts is a mistake. It is also a mistake to attempt to harmonise the seven day story with an evolutionary unfolding of creation by extending each day into an epoch. This only perpetuates the mistake of treating Genesis 1-3 as an historical source in the modern sense. The account of creation in Genesis 1 seeks to communicate truth. But it is biblical truth, the truth about God, about God's creatures, and about human beings before God.

A second misreading would be to see it providing any kind of scientific information for today. The world-view of Genesis was one that was common to the peoples of their area. They saw the inhabited word as emerging from and as completely surrounded by the 'abyss' (Gen 1:2) of the primeval ocean. God creates a huge dome to separate the waters above us from the waters below (Gen 1:7). This is the 'firmament', a concave plate, creating a hollow space between the upper and lower waters. The Hebrew word means 'something hammered out flat'. At certain times, gates in the dome of the sky are opened, so that the water above us can come through to fall as rain on the ground. The water below us springs forth as fountains and rivers. God gathers the waters below the dome together as seas, so that the dry land appears safely separated from the seas (Gen 1:9). God attaches to the dome of the sky the great lights, the sun for the day, and the moon and stars for the night (Gen 1:14-19). This kind of cosmology is not authoritative in any way for a Christian today. We rightly turn to sciences like astronomy

---

6. There are a number of very different cosmogonies in the Bible, in Proverbs, Job, Psalms and Second Isaiah as well as the Genesis accounts.

Evolution and the Christian God

and cosmology to learn about the emergence and structure of the universe. We turn to the Bible not for cosmology but for religious insights into God as Creator.

What, then, are these religious insights? Of course to simply list a number of religious concepts cannot do full justice to what is presented in a symbolic, narrative form. But granted this, it is clear that the author of Genesis 1 aims to teach that creation is the free act of the one God of the covenant. This one saving God is the creator of all things. Nothing in nature is God. God is preexistent and transcendent over all creatures. But this God relates closely to creation, and takes delight in all creatures. All of creation is good. It is God's blessing that enables creation to be fruitful and fecund. Human beings, male and female, are made in the image of God. They are created as social beings, able to communicate with God. The Sabbath is a sign of God's care for human beings and for all creation, and functions as a promise of future joy and peace in God. Genesis 1 is a picture of what God intends for creation. Because of this it is also an eschatological statement, a promise of the future. The stories of human sin that follow will not annul this promise.

These religious insights constitute a rich theological heritage for Christian theology. They remain a fundamental resource for a theology of creation. In themselves they are not in any way in conflict with evolutionary biology. It is entirely possible to hold to these religious insights and at the same time to hold to the evolution of life on earth over the last 3.8 billion years. We should not look to Genesis for scientific ideas about the history of the early universe or the emergence of life. But a Christian can read these narratives for religious truth that will be essential components of a contemporary theology of creation. There is every reason for a Christian of today to embrace *both* the theological teachings of Genesis and the theory of evolution.

2. God's purposes and the randomness of evolution

As John Hedley Brooke has shown, science and natural theology were closely linked in the first half of the nineteenth century.[7] Many scientists were from the ranks of the clergy, and much scientific work was thought of as revealing the hand of the divine Designer. In 1802, William Paley had published his *Natural Theology*, a widely read work that influenced many scientists. Paley had been struck by the exquisite design that is apparent in the structure of living organisms. It appeared obvious to him that the way the eye is designed for vision, for example, could only be the work of a divine Designer.

This popular and successful line of thought was radically challenged by Darwinian theory, which was able to explain the emergence of the eye and other organisms by means of natural selection without recourse to a Designer. Neo-Darwinian biology not only challenges the need for a divine Designer, but also gives a large role to chance. It sees randomness as built into evolutionary process. The genetic mutations that are the source of novelty in natural selection arise at random. Some of these mutations are beneficial, but many are harmful. Without these random mutations, evolution through natural selection could not occur because they are the source of the variations passed on to other generations. Although evolution is dependent upon random mutation, evolution itself is not merely random, because natural selection functions positively to preserve what is useful for adaptation to an environment and to eliminate what is not useful. It is mutation and natural selection working together that produce something as beautiful as a lorikeet and as complex as the human brain. It is chance and lawfulness working together that have brought forth the exuberant diversity of life on earth. Chance runs through the whole pattern of life on earth, not only in the randomness of mutations, but also in one-off accidental and contingent events such as the collision of a meteor with the earth sixty-five million years ago that led to the extinction of the dinosaurs and created the space for mammals to flourish.

---

7. John Hedley Brooke, *Science and Religion: Some Historical Perspectives* (Cambridge: Cambridge University Press, 1991), 192-225.

Evolution and the Christian God

How does this understanding of the role of chance and contingency in evolution relate to the Christian view of God as the Creator? Jacques Monod, in his *Chance and Necessity*, argued that since evolution is grounded in 'pure chance', there is no longer any point in talk of purpose or meaning in the universe.[8] This amounted to a powerful attack on the Christian view of God as Creator, an attack that has been carried on in a different form in recent years by writers on evolution like the biologist Richard Dawkins and the philosopher Daniel Dennett.[9] Stephen Jay Gould denies that progress or increased complexity is characteristic of life as a whole. He argues that if we could play the tape of life again, the vast majority of replays would not produce a creature with self-consciousness. He claims that we human beings are 'glorious accidents of an unpredictable process'.[10] The eminent zoologist Ernst Mayr has said that science has not found any teleological mechanisms that would justify us talking about overall purpose.[11] He insists that neo-Darwinian science is the ultimate explanation of life.[12]

It is important to note that there are also many biologists who believe in a creative God, and find in the abundance and diversity of life, and in biological evolution itself, reason for religious awe and thanksgiving. Australian biologist Charles

---

8. Jacques Monod, *Chance and Necessity* (London: Collins, 1972).
9. See for example, Richard Dawkins, *The Blind Watchmaker* (Harlow: Longman, 1986); Daniel C Dennett, *Darwin's Dangerous Idea* (London: Allen Lane, 1995).
10. Stephen Jay Gould, *Life's Grandeur: The Spread of Excellence from Plato to Darwin* (London: Vintage Books, 1996), 216. Gould does not deny increased complexity in some species, but sees this as restricted to only a few species, and as an incidental effect rather than as the intended result of evolutionary change (197). See also his *Wonderful Life: The Burgess Shale and the Nature of History* (London: Penguin Books, 1989), 35.
11. Ernst Mayr, 'Evolution', *Scientific American*, 134 (September 1978): 50.
12. Ernst Mayr, *This Is Biology* (Cambridge, Mass: Harvard University Press, 1997), 64.

Birch, for example, has written a number of books defending the idea of divine purpose at work in the universe.[13]

It is important too, to note that comments of scientists on the existence or non-existence of a Creator, or on the existence or non-existence of purpose and meaning in the universe, take them beyond scientific competency into the fields of philosophy and theology.

In my view, comments such as those of Ernst Mayr need to be taken seriously by theology. But they need to be understood as statements about biology. What a biologist can say legitimately, as a biologist, is that the appearance of design in the eye or the brain can now be explained satisfactorily by the theory of natural selection, and there is no evidence from biology that an external divine designer is needed. This claim in itself can coexist with belief in God or with a rejection of God. I think it is a mistake for theologians to insist that divine action is needed for a satisfactory *biological* explanation. This runs the risk of falling into the trap of the old 'god-of-the-gaps'. If someone like Ernst Mayr concludes that biology does not give empirical evidence of purpose at work in the universe, then, as a theologian I am inclined to accept his conclusion—at least until it is challenged by further *biological* evidence.

But as a theologian, I would need to note that, supposing Mayr is right about the state of biological evidence, this does not rule out a theological principle of purpose. It is quite possible to think theologically of God as working purposefully in the universe, through processes such as random mutation and natural selection that, when investigated empirically, do not reveal purpose at all. Theologically, it is possible to think of God's purposes being achieved through what appears to empirical biology to be without purpose.

Darwinian biology has succeeded in discrediting William Paley's kind of natural theology. By providing an adequate

---

13. Charles Birch, *On Purpose* (Sydney: UNSW Press, 1990); *Biology and the Riddle of Life* (Sydney: UNSW Press, 1999).

Evolution and the Christian God

explanation for the exquisite functional design of the eye, it successfully challenges Paley's argument that such examples of functional design require the existence of a divine Designer. Biology can no longer be used in this way as an apologetic for Christian faith. I consider this as a helpful gain for theology. In a recent important book on evolution, the theologian John Haught points out that the theory of natural selection can actually be considered as 'Darwin's gift to theology', enabling theology to move beyond fruitless design arguments towards a deeper evolutionary theology.[14]

Such an evolutionary theology will bring together the theory of biological evolution through random mutation and natural selection with a theology that is grounded in God's self-giving to the world in Jesus Christ. It will embrace both the contingency of evolution and the biblical teaching of God's faithful love for human beings and all of creation. It will see God's purposes as being worked out, at least in part, through the processes of random mutation and natural selection.

According to Christian theology, God is as close to us as we are to ourselves and yet remains absolute mystery. God's present is radically relational. God is not a solitary individual but a trinitarian communion. God's very being is communion and God acts in creation in a trinitarian way. The ancient Christian tradition sees God as creating and renewing all things through the Word and in the Spirit. It sees God as present to all things in the Spirit, who is the ecstasis of the divine communion. The Spirit is the presence of God at the heart of things, the bringer of communion to all creatures. It is this Holy Spirit that we confess in the creed as the 'Life-Giver'. When God acts in creation, this is always the action of the Creator Spirit, in union with the Word and the one who is the source of all being.

This presence of God in the Spirit transcends all our concepts and all language. When we speak of 'divine action' or of God

---

14. John F Haught, *God after Darwin: a Theology of Evolution* (Boulder, Colorado: Westview Press, 2000), 23-56.

achieving 'purposes', we use language in an analogical fashion. This kind of language is based on our own human experience of acting and influencing events, and projecting this onto God. But, if we are theologically aware, we will allow that God does *not* act and influence events in a human way but in a way that radically transcends all human notions of agency. We influence events through *intervention* of one kind or another—by word or action we help to shape the outcome we desire. An interventionist theology reduces God to being one cause amid other inner-worldly causes and effects. When we speak of God acting purposefully, we naturally tend to think of God intervening in creation. But the Spirit of God cannot be reduced to one player among others, a cause among other causes in the world. The theological tradition has resisted seeing God's action as an intervention in the natural world.

We need to be able to think of God acting in the universe purposefully, but we need to acknowledge that this is not a kind of action that we can easily pin down or imagine. We cannot locate the 'place' of this kind of action. God's action remains ever mysterious to us. What we know from Christian theology is that God can act to achieve purposes through random and contingent events. The events of Jesus' life were contingent upon both human freedom and the constraints of the natural world. The death of Jesus on the cross, for example, was contingent on many human factors, including Jesus' choice to go to Jerusalem, his prophetic action in the temple, the betrayal of Judas, his condemnation by religious authorities, and the decision of Pilate to execute him. It was probably contingent on the fact that it was the time of the Passover. The death was itself a cruel and barbaric act. Yet, for Christian theology, this event in all its contingency and brutal negativity is transformed by God's 'action' and 'purpose' so that it becomes in a mysterious way the source of life for all creatures. Clearly Christian faith is based upon the fact that God can act in and through unpredictable and contingent events—even contingent events that are in themselves evil.

Evolution and the Christian God

In a similar way the Christian doctrine of grace has always affirmed that graced human beings remain fully free to be themselves. Grace does not destroy human freedom but allows human freedom to flourish. In grace the Spirit of God seems to be able to achieve 'purposes' in and through the contingency of real human freedom. We are not less free when we are Spirit-led, but most fully free and most fully autonomous.

Karl Rahner has put forward as a fundamental axiom of theology that human freedom and dependence on God grow in equal not inverse proportions.[15] The intuitive or 'common sense' approach would be that the more we are dependent on God the less free we are. Rahner insists that the more we grow in authentic freedom the more we are dependent on God and in authentic union with God. This is what DM Baillie has called the *paradox of grace*—the experience that 'never is human action more truly and fully personal, never is the agent more perfectly free' than in those moments when Christians know that whatever good was in them was not theirs but God's.[16] What Rahner and Baillie point to is the Christian conviction that God's purposes can be accomplished through the contingency of human freedom. In a similar way, I believe, we need to acknowledge that God can work artfully and purposefully in the unpredictability and contingency of nature—the 'freedom' of nature.

This theology of divine purposes being achieved in and through the contingencies of the life and death of Jesus, and in and through the contingency of human free acts in the life of grace, suggest that in a theology of creation, God can be understood as achieving 'purposes' in the contingencies of natural processes. What we know of divine action in the theology of incarnation and of grace is that God 'acts' through

---

15. See, for example, Karl Rahner, 'Freedom. II. Theological', in Karl Rahner, editor, *Encyclopedia of Theology: A Concise Sacramentum Mundi* (London: Burns and Oates, 1975), 545.
16. DM Baillie, *God Was in Christ* (London: Faber and Faber, 1956), 114.

Denis Edwards

contingent events. It should not be surprising to find that God's action in creation occurs in and through contingent and chancy events.

The God of evolution is therefore to be understood as a God involved creatively in an open-ended process, which involves both randomness and lawfulness. This kind of open process is intrinsic to the universe we inhabit. DJ Bartholomew has urged that chance 'plays a constructive role in creating a richer environment than would otherwise be possible'. He sees chance as giving the Creator advantages that it is difficult to envisage being obtained in any other way. He believes that there is every reason to think that a Creator who wished to achieve certain ends, such as intelligent creatures, might choose to reach those ends by means of random but creative processes.[17] In a similar way, Arthur Peacocke sees the role of chance as 'simply what is required if all the potentialities of the universe, especially for life are to be elicited effectively'.[18] In the world as we know it, biological life is inconceivable apart from evolution through random mutation.

Peacocke offers two analogies for divine interaction with the world that I find helpful. First, he suggests the idea of '*top-down*' causation as a way of thinking about God's action with regard to the world. This language comes from the sciences of complexity, where it has become recognised that alongside 'bottom-up' scientific explanations there is need to take into account the state of the system-as-a-whole. This state of a system-as-a-whole can be regarded as a causal factor influencing events at the lower level. On this basis, Peacocke suggests that the world might be considered as a total system and God's relationship with it understood according to the model of top-down causation.[19] This enables us to think of God

---

17. David J Bartholomew, *God of Chance* (London: SCM, 1984), 97-98.
18. Arthur Peacocke, *Theology for a Scientific Age*, Enlarged Edition (London: SCM, 1993), 120.
19. *Ibid*, 157-160.

interacting with the world as a whole, is such a way that God makes a difference in our world, but not in a way that is contrary to the regularities and laws of the natural world. Peacocke points out that there is a parallel with the top-down way that human thought influence bodies: 'God would be regarded as exerting continuously top-down causative influences on the world-as-a–whole in a way analogous to that whereby we in our thinking can exert effects on our bodies in a "top-down" manner'.[20] The little word 'analogous' is important. It is fundamental to keep stressing that God's presence and action radically transcends all examples of top-down or whole-part causation in this world. But, granted this, I find this image a helpful one.

The second analogy that I find illuminating is Peacocke's image of God as an explorer and *improvisor* in creation. He sees God as 'involved in explorations of the many kinds of unfulfilled potentialities of the universe', potentialities which have been given to the universe by God. For Peacocke, God at work in creation is like a composer able to unfold all the hidden possibilities of a musical theme. He sees God as the 'Improviser of unsurpassed ingenuity'. God at work in creation is like Bach improvising on a theme at the keyboard. He notes that introducing improvisation into this image of God as composer 'incorporates that element of open adaptability which any model of God's relation to a partly non-deterministic world should, however inadequately, represent'.[21]

3. The suffering of creation and divine self-limitation in love
Christian theology celebrates the diversity and beauty of life as a gift of the Creator. But it cannot afford to take a romantic view that obscures pain and death. Great White sharks eat seals. Cats drag home small birds. Creatures prey on other creatures in

---

20. *Ibid*, 161.
21. *Ibid*, 175.

order to live. Predation and death are part of the pattern of biological life.

Pain and death, and all that might be called natural evil as opposed to human evil, have always been a problem for Christian theology. This theological issue is not the product of evolutionary biology. But a biological worldview brings the issue of natural evil into sharp focus. It points to the unthinkably long history of life on earth stretching back for 3.8 billion years, with all this means in terms of the lives and deaths of countless creatures. It makes us aware of massive extinctions of life, including the event of 248 million years ago, which destroyed more than ninety per cent of marine species, and that of sixty-five million years, which wiped out the dinosaurs along with sixty-five per cent of the earth's species. Biology teaches us about the random mutations that provide the novelty for evolutionary change, but which also bring suffering and death. It points to extraordinary examples of symbiosis and cooperation, but it also indicates the place of competition and the struggle for existence in the story of evolution. Evolutionary biology insists on the fundamental place of death—without death there would be no series of generations and therefore no evolution. Ursula Goodenough writes: 'Death is the price paid to have trees and clams and birds and grasshoppers, and death is the price paid to have human consciousness'.[22]

All of this challenges theologians to move beyond romantic views of creation to a deeper theology of creation. Christian theology has no satisfying theoretical answer to the issue of pain and death in nature. It simply has to face the fact that this is the way things are in this finite, limited, bodily world. Christian theology can only bear witness to the death and resurrection of Jesus. Here divine love is revealed as unthinkable compassion. The cross reveals a God who enters into the pain of the world, who suffers with suffering creation.

---

22. Ursula Goodenough, *The Sacred Depths of Nature* (New York/Oxford: Oxford University Press, 1998), 151.

Evolution and the Christian God

In the resurrection, through the life-giving Spirit new life is promised and in some way already given, and those who live in the power of resurrection life are convinced that forgiveness, liberation and new creation have already taken hold at the heart of this world.

The cross reveals something about the nature of divine action and divine power. It tells us that the divine capacity for love involves not only boundless generosity but also an incomprehensible vulnerability and free self-limitation. It seems to me that we Christians have often got this issue wrong. When we think of divine power at work in creation, we can all too easily base our understanding of divine power not on the cross and resurrection of Jesus, but on the way human tyrants exercise oppressive and dominating power. Of course we rightly understand God as the one whose power creates the universe and will bring it liberation and fulfilment. We rightly confess in the Creed that God is 'almighty'. But the crucial issue concerns what kind of might or power we attribute to God. It is easy to assume that God is *absolutely* unlimited in power, so that such a God could do anything, without qualifications. Such a God could arbitrarily overrule human freedom and the laws of nature. This attitude to divine power is revealed when, in the face of tragedy or difficulty, people say: 'Why is God doing this to me?' It is revealed also when the response is given: 'It's God's will'. A God with this kind of power could intervene to stop a tidal wave or to ensure that a dangerous mutation did not occur. If one has this view of God's omnipotence, then it is natural enough to see God as capricious and arbitrary in not acting to stop evil. But where does this view of divine power come from? I believe that it comes not from what we find at the heart of the Christian Scriptures, but from transferring to God the kind of notions of sovereignty and power that appear in the worst excesses of human emperors, kings and dictators. God is understood as the great lordly individual able to do and to command anything no matter how arbitrary or excessive.

Denis Edwards

If we turn to the Christian scriptures we find a different view of divine power. In Mark's gospel, for example, we find that the messianic identity of Jesus is radically connected to the cross. The gospel presents Jesus as messianic Son of God, but it will not allow us to understand Jesus as a messiah who wields dominating power. His true identity is revealed only in his death on the cross. It is this vulnerable, defeated and humiliated one that is proclaimed as messianic Son of God by the centurion (Mark 15:39). It is this crucified one who has been raised up to go before the disciples into Galilee (Mark 16:6). In Philippians, Christ Jesus is understood as one 'who did not think of equality with God something to be exploited, but emptied himself, taking the form of a slave'. (Phil 2:7) This self-emptying (*kenosis*) describes the pattern of the way that God is among us in Christ, culminating in the cross. In First Corinthians, Paul describes this whole kenotic pattern in response to claims about who is superior in wisdom:

> For Jews demand signs and Greeks desire wisdom, but we proclaim Christ crucified, a stumbling block to Jews and foolishness to Gentiles, but to those who are the called, both Jews and Greeks, Christ the power of God and the wisdom of God. For God's foolishness is wiser than human wisdom, and God's weakness is stronger than human strength (1 Cor 1:22-25).

Christ crucified now defines the nature of divine power. Divine power is now understood as the power of love. This is the 'foolishness' that is radically beyond all human wisdom. These early Christian texts all insist that it is the kind of power we find revealed in Jesus that is to be lived out in the Christian community. Paul tells the Philippians that Christian life will involve taking on 'the same mind that was in Christ Jesus' (Phil 2:5). He tells the Corinthians that he did not approach them in a superior way: 'I decided to know nothing among you except

Evolution and the Christian God

Jesus Christ, and him crucified' (1 Cor 2:1). Mark tells his community that there is to be absolutely no 'lording' it over others in the Christian community (Mark 10:42-45). All forms of dominating power are forbidden among those who follow the way of Jesus. The pattern of kenotic love must shape every expression of power in the Christian community.

If God is to be understood as true and faithful, then the Christic pattern of vulnerable and self-limiting love can be understood to govern not just the story of Jesus and the church, but also God's creative presence to all creatures in the Spirit. God's power is revealed as the power to love. Such a capacity for free self-limitation in love is, then, not to be understood as a diminishment of God or a loss of transcendence, but as the real expression of divine transcendence. God's transcendence is revealed as the capacity to love in a way that is beyond all human comprehension, the capacity to give oneself in love in a way beyond human possibilities.

The cross and resurrection reveal the nature of divine power as the power to love. It is because God is omnipotent in love that God can freely enter into the vulnerability of loving. Love involves not only self-communication to the other, but also self-limitation. The lover needs to 'make space' for the other and allow the other to exist in difference and distinction. The lover allows the beloved to affect him or her, and can suffer with the beloved. This involves a form of self-limitation, a letting be of the other, a making room for the other. The God revealed in the cross is a God who is capable of freely entering into the self-limitation and vulnerability of love, without compromising the proper autonomy of the creature on the one hand or divine transcendence on the other.

God's self-revelation of the cross suggests a Creator who relates in compassionate, self-limiting love with creatures. Such a love would respect each creature's proper nature and autonomy. It points to a God who suffers with creation not out of necessity and not out of imperfection, but in the active freedom of the divine love. God suffers with creation not to glorify suffering but in order to bring liberation and healing. This divine power is at work in every creature, in every aspect of ongoing creation, in every human event, even in places like Golgotha—above all there. But it is a power defined by love, a power that makes room for creatures to be their finite creaturely selves. It makes room for men and women to be authentically free and it makes room for other creatures to be themselves. It is a power that is self-limited because in the freedom of love it respects the otherness and autonomy of physical processes.

This suggests that the Creator Spirit, because of divine free self-limitation which respects the otherness of creatures, may not be free to overturn the proper unfolding and emergence of creation, and may well be committed to respecting the proper autonomy and independence of all things. In Romans 8, Paul sees the creation as in childbirth: 'For we know that the entire creation has been groaning together in the pangs of childbirth up till now' (8:22). He writes of creation groaning in the suffering of childbirth, human beings groaning in anticipation of God's future, and the Spirit groaning deep within them. It seems entirely appropriate to build on Paul's thought to see the Spirit as the companion of creation in its travail and as the midwife of new creation. All through the ages, the Christian community has continued to cry out 'Come Holy Spirit! Come Creator Spirit!' It continues to pray: 'Send out your Spirit and renew the face of the earth'. The transformation that we call for is imaged in biblical texts as an overcoming of the struggle, as *shalom*, as entering into the Sabbath of God. It is imaged as a liberation from both predation and death. This needs to be understood as an image for what is beyond imagining—the future participation of all creatures in the dynamism of the divine life.

The evolutionary process involves struggle, pain and death. Jay McDaniel has said that the appropriate Christian response is 'to be honest and open about the violence around us and within us, not hiding from it; trustful that the very Heart of the universe suffers with each and every living being that suffers;

Evolution and the Christian God

and inspired toward a non-violent way of living that shares with the world the non-violence of God'.[23] The Spirit is present with each and every creature, as cosufferer and companion. The Spirit groans with each creature, feeling its pain, enabling it to have its own integrity, and longing for its completion and fullness of life in God. For us human beings, to experience the Spirit in this way can lead to a radical compassion for other creatures.

4. Conclusion

There is, of course, much more that might be said about a theological approach to evolution.[24] Here I have tried to deal briefly with three fundamental issues. First, I have suggested that when the Genesis creation accounts are read in terms of their literary form, they are not a source for scientific knowledge, but do convey profound religious insights. These insights are congruent with an evolutionary view of the world. Second, I have argued that it is entirely possible to accept the claims of biologists that the evolution of life involves random mutations and contingent events, and at the same time to hold to the Christian view that God is creating purposefully. God can achieve purposes in a way that is beyond human comprehension, in and through events that from a human perspective appear entirely contingent. Third, I have argued that the painful costs associated with the evolution of life can be better faced with a theology which accepts that God is freely self-limiting in creation, respecting each creature's integrity, and the unfolding of natural processes. God is present to each creature in the Spirit, loving each one, delighting in each one,

---

23. Jay McDaniel, *With Roots and Wings: Christianity in an Age of Ecology and Dialogue* (Maryknoll: Orbis Books, 1995), 53.
24. I have tried to offer a more developed evolutionary theology in *The God of Evolution: A Trinitarian Theology* (New York: Paulist Press, 2000). See also John Haught's *God after Darwin: A Theology of Evolution* and Arthur Peacocke's *Theology for a Scientific Age*.

Denis Edwards

suffering with them in their pain, relating to each in a way that respects each one's uniqueness, and drawing all into the divine life in a healing and liberating new creation.

Historiographical Resources for Teaching Religion and Science

Complied by Peter Hess, Center for Theology and the Natural Sciences

Barbour, Ian G, *Religion and Science: Historical and Contemporary Issues* (San Francisco: HarperCollins, 1997). [Part One discusses history since the seventeenth century.]

Barry, James, 1955-. *Measures of science: theological and technological impulses in early modern thought* (Evanston, Ill: Northwestern University Press, 1996).

Bono, James J, *The Word of God and the Languages of Man: Interpreting Nature in Early Modern Science and Medicine* (Madison: University of Wisconsin Press, 1995).

Brooke, John H, *Science and Religion: Some Historical Perspectives* (Cambridge University Press, 1991). [Brooke conveys with great detail and sophistication the complexities of the science-religion relationship from the Renaissance to the nineteenth century.]

Brooke, John H, and Cantor, Geoffrey, *Reconstructing Nature: The Engagement of Science and Religion* (Edinburgh, T&T Clark, 1998).

Brooke, John Hedley, Van Der Meer, Jitse M, and Osler, Margaret J, editors, *Science in Theistic Contexts: Cognitive Dimensions* (University of Chicago Press, 2001).

Historiographical Resources

Cosslett, Tess, *Science and Religion in the Nineteenth Century* (Cambridge, 1984). [A valuable collection of primary sources for the period.]

Dobbs, Betty Jo Teeter, and Jacob, Margaret C, *Newton and the culture of Newtonianism* (Atlantic Highlands, NJ: Humanities Press, 1995).

Ferngren, Gary B, editor, *The History of Science and Religion in the Western Tradition: An Encyclopedia* (Garland, 2000). [An indispensable reference work of 106 substantial essays treating particular figures, central themes, religious traditions, and the branches of science.]

Force, James E, *Essays on the Context, Nature, and Influence of Isaac Newton's Theology* (Dordrecht ; Boston : Kluwer Academic Publishers, 1990).

Force, James E, and Popkin, Richard H, editors, *Newton and Religion: Context, Nature, and Influence* (Archives Internationales D'Histoire Des Idees, 161.) Dordrecht, Boston: Kluwer,1999.

Glacken, Clarence, *Traces on the Rhodian Shore: Nature and Culture in Western Thought from Ancient Times to the End of the Eighteenth Century* (Berkeley: University of California Press, 1967). [A classic, with much useful interpretation *passim* of 'nature' in theological context throughout.]

Harrison, Peter, *The Bible, Protestantism, and the Rise of Modern Science* (Cambridge, 1998). [An engaging work examining the role of the Bible in the emergence of Western science.]

Knight, David M, *Natural Science Books in English, 1600-1900* (London: Batsford, 1989). [This nicely illustrated work

Historiographical Resources

places early scientific writing in its historical, theological, and cultural contexts.]

Lindberg, David C, *The Beginnings of Western Science: The European Scientific Tradition in Philosophical, Religious, and Institutional Context, 600 BC to AD 1450* (Chicago: University of Chicago Press, 1992). [A fine synthetic work that pays careful attention to religious influences.]

Lindberg, David C, and Numbers, Ronald L, editors, *God and Nature: Historical Essays on the Encounter between Christianity and Science* (Berkeley: University of California Press, 1984). Eighteen insightful essays that definitively demolished the 'Warfare Myth'.]

Lindberg, David C, and Numbers, Ronald L, editors, *Science and the Christian Tradition: Twelve Case Histories* (Chicago, 2001). [Promises to be an excellent pedagogical tool.]

Livingstone, David, DG Hart, and Mark A Noll, editors, *Evangelicals and Science in Historical Perspective* (Oxford, 1999).

Manuel, Frank E, *The Religion of Isaac Newton* (Oxford: Clarendon Press, 1974).

Olson, Richard, *Science Deified and Science Defied: The Historical Significance of Science in Western Culture*, volumes 1 & 2 (Berkeley: University of California Press, 1982, 1990). [An excellent broad survey of the reciprocal influences of science and culture upon each other.]

Osler, Margaret J, editor, *Rethinking the Scientific Revolution* (Cambridge, 2000). [Redefines this pivotal period, showing the intricate relationships between religion and science in theory and practice.]

Historiographical Resources

Prest, John, *The Garden of Eden: The Botanic Garden and the Recreation of Paradise* (New Haven: Yale University Press, 1981). [A delightfully illustrated book, tracing the foundation of botanic gardens to the Early Modern desire to reassemble the elements of creation scattered by the Flood.]

Rudwick, Martin JS, *Scenes from Deep Time: Early Pictorial Representations of the Prehistoric World* (Chicago: The University of Chicago Press, 1992). [A beautifully illustrated work, that traces significant influences—including religious ones—on the development of the iconographic tradition (1500-1900) of depicting deep geological time.]

Southgate, Christopher, *God Humanity and the Cosmos* (Harrisburg, PA: Trinity Press, 1999). [This textbook nicely integrates historiographical considerations into every chapter.]

Welch, Claude, 'Dispelling Some Myths about the Split between Theology and Science in the Nineteenth Century', in W Mark Richardson and Wesley J Wildman, editors, *Religion and Science: History, Method, Dialogue* (New York & London: Routledge, 1996), 29-40.

Brief Bibliography in Science and Religion

The Center for Theology and the Natural Sciences

This is a working bibliography, rather than a comprehensive list of works, central to the field of science and religion. Included are classic broad treatments of the field, as well as recent works on more specific topics.

Basic Level

Ashbrook, James B, and Carol R Albright, *The Humanizing Brain: Where Religion and Neuroscience Meet* (Pilgrim, 1997) ISBN: 0829812008.

Brockelman, Paul, *Cosmology and Creation: The Spiritual Significance of Contemporary Cosmology* (London: Oxford University, 1999) ISBN: 0195119908.

Cole-Turner, Ronald, *Human Cloning: Religious Responses* (Westminster, 1997) ISBN: 0664257712.

Davies, PCW, *God and the New Physics* (Dent, 1983) ISBN: 0671528068.

———, *The Mind of God: the Scientific Basis for a Rational World* (Touchstone, 1993) ISBN: 0671797182.

Edwards, Denis, *Jesus and the Cosmos* (Mahwah, NJ: Paulist, 1992) ISBN: 0809132214.

CTNS Bibliography

———, *The God of Evolution: A Trinitarian Theology* (Mahwah, NJ: Paulist, 1999) ISBN: 0809138549.

Ellis, George FR, and Peter H Collins, *Before the Beginning: Cosmology Explained* (Marion Boyars, 1993) ISBN: 0714529702.

Ferris, Timothy, *Coming of Age in the Milky Way* (Anchor, 1989) ISBN: 0385263260.

Matt, Daniel, *God and the Big Bang: Discovering Harmony between Science and Spirituality* (Jewish Lights, 1996) ISBN: 1879045893.

Peters, Ted, editor, *Cosmos as Creation: Theology and Science in Consonance* (Abingdon, 1989) ISBN: 0687096553.

Polkinghorne, John, *Science and Providence: God's Interaction with the World* (Society For Promoting Christian Knowledge, 1989) ISBN: 0281043981.

———, *Science and Theology: An Introduction* (Minneapolis: Fortress, 1999) ISBN: 0800631536.

———, *Belief in God in an Age of Science* (New Haven, CT: Yale University, 1998) ISBN: 0300080034.

Sobosan, Jeffrey G, *The Turn of the Millennium: An Agenda for Christian Religion in an Age of Science* (Cleveland: Pilgrim, 1996) ISBN: 0829810838.

CTNS Bibliography

Southgate, Christopher, editor, *God, Humanity and the Cosmos: A Textbook in Science and Religion* (Harrisburg, PA: Trinity, 1999) ISBN: 1563382881.

Templeton, John M, editor, *How Large Is God? The Voices of Scientists and Theologians* (Templeton Foundation, 1997) ISBN: 1890151017.

Tilby, Angela, *Soul: God, Self, and the New Cosmology* (Doubleday, 1992) ISBN: 0385471254.

Wertheim, Margaret, *Pythagoras' Trousers: God, Physics, and the Gender Wars* (WW Norton & Co, 1997) ISBN: 0393317242.

Intermediate Level

Barbour, Ian, *Issues in Science and Religion* (Prentice Hall, 1966) ISBN: 006135664.

_____, *Myths, Models and Paradigms* (Harper & Row, 1974) ISBN: 0060603887.

_____, *Ethics in an Age of Technology: The Gifford Lectures,* volume 2 (HarperCollins, 1992) ISBN: 0060609354.

_____, *Religion and Science: Historical and Contemporary Issues,* Revised Edition (HarperCollins, 1997) ISBN: 006060939.

Barr, James, *Biblical Faith and Natural Theology: The Gifford Lectures for 1991* (London: Oxford Univiersity, 1995) ISBN: 0198263767.

CTNS Bibliography

Bartholomew, D, *God of Chance* (Harrisburg, PA: Trinity, 1984) ISBN: 0334020301.

Birch, Charles, William Eakin and Jay B McDaniel, editors, *The Liberation of Life* (Orbis Books, 1990) ISBN: 0883446898.

Brooke, John Hedley, *Science and Religion: Some Historical Perspectives* (Cambridge, MA: Cambridge University Press, 1991) ISBN: 0521283774.

Brown, Warren, and Murphy, Nancey, editors, *Whatever Happened to the Soul? Scientific and Theological Portraits of Human Nature* (Minneapolis: Fortress, 1998) ISBN: 0800631412.

Cole-Turner, Ronald, and Waters, Brent, *Pastoral Genetics: Theology and Care at the Beginning of Life* (Westminster/John Knox, 1996) ISBN: 0829810773.

Dillenberger, John, *Protestant Thought and Natural Science* (University of Notre Dame Press, 1989) ISBN: 0268015759.

Durant, John, editor, *Darwinism and Divinity* (Blackwell, 1986) ISBN: 063115101x.

Fagg, Lawrence W, *The Becoming of Time* (Scholars, 1995) ISBN: 0788500597.

CTNS Bibliography

Frye, Roland, M *Is God a Creationist? The Religious Case against Creation Science* (Scribner, 1983) ISBN: 0684180448.

Gilkey, Langdon, *Creationism on Trial* (Winston, 1986) ISBN: 0866837809.

_____, *Nature, Reality and the Sacred* (Minneapolis: Fortress, 1994) ISBN: 0800627547.

Haught, John F, *Science and Religion: From Conflict to Conversation* (Paulist, 1995) ISBN: 0809136066.

Hefner, Philip, *The Human Factor: Evolution, Culture, and Religion* (Minneapolis: Fortress, 1993) ISBN: 080062579x.

Huchingson, James E, *Religion and the Natural Sciences: the Range of Engagement* (Hbj College, 1997) ISBN: 0030522536.

Jeeves, Malcolm, *Human Nature at the Millennium: Reflections on the Integration of Psychology and Christianity* (Inter-Varsity Press, 1997) ISBN: 0851114512.

Klaaren, Eugene, *Religious Origins of Modern Science* (Wm B Eerdmans, 1978) ISBN: 0802816835.

Lindberg, David C, *The Beginnings of Western Science* (Chicago: University of Chicago Press, 1992) ISBN: 0226482316.

Lindberg, David C, and Numbers, Ronald L, editors, *God & Nature: Historical Essays on the Encounter Between Christianity*

CTNS Bibliography

and Science (Berkeley, CA: University of California Press 1986) ISBN: 0520056922.

Loder, James E, and W Jim Neidhardt, *The Knight's Move* (Helmers & Howard, 1992) ISBN: 0939443252.

McFague, Sallie, *Metaphorical Theology* (Minneapolis: Fortress Press, 1997) ISBN: 0800616871.

_____, *The Body of God* (Fortress Press, 1993) ISBN: 0800627350.

Matthews, Clifford N, and Varghese, Roy Abraham, *Cosmic Beginnings and Human Ends: Where Science and Religion Meet* (Open Court, 1994) ISBN: 0812692691.

Moltmann, Jürgen, *God in Creation* (Minneapolis: Fortress Press, 1993) ISBN: 0800628233.

Murphy, Nancey, *Theology and Science: A Radical Reformation Perspective* (Kitchener, Ontario: Pandora, 1997) ISBN: 0969876246.

Numbers, Ronald, *Darwinism Comes to America* (Cambridge, MA: Harvard University Press, 1998) ISBN: 0674193113.

Peacocke, Arthur, editor, *The Sciences and Theology in the Twentieth Century*

CTNS Bibliography

———, *Theology for a Scientific Age: Being and Becoming—Natural, Divine and Human* (Minneapolis: Fortress, 1993) ISBN: 0800627598.

Peters, Ted, *God as Trinity* (Westminster John Knox, 1993) ISBN: 0664254020.

———, *Science and Theology: The New Consonance* (Westview, 1998) ISBN: 0813332583.

———, *Playing God? Genetic Determinism and Human Freedom* (New York and London: Routledge, 1997) ISBN: 0415915228.

Polkinghorne, John, *Faith of a Physicist: Reflections of a Bottom-Up Thinker* (Minneapolis: Fortress, 1996) ISBN: 0800629701.

Rolston, Holmes, III, *Environmental Ethics* (Temple University Press, 1989) ISBN: 0877226288.

———, *Science and Religion, a Critical Survey* (Temple University Press, 1987) ISBN: 0877224374.

Soskice, Janet, *Metaphor and Religious Language* (Oxford University Press, 1987) ISBN: 0198249829.

Santmire, H, Paul, *The Travail of Nature* (Minneapolis: Augsburg Fortress) ISBN: 0800618068.

CTNS Bibliography

Stannard, Russell, *Grounds for Reasonable Belief* (Scottish Academic, 1989) ISBN: 0707305810.

Teilhard de Chardin, Pierre, *The Phenomenon of Man* (Harper Collins, 1980) ISBN: 006090495X.

Theissen, Gerd, *Biblical Faith: An Evolutionary Approach* (Minneapolis: Fortress, 1985) ISBN: 0800618424.

Tracy, Thomas F, *The God Who Acts* (Pennsylvania State University Press, 1994) ISBN: 0271010398.

Van Huyssteen, Wentzel, *Theology and the Justification of Faith* (Wm B Eerdmans, 1989) ISBN: 0802803660.

_____, *Duet or Duel? Theology and Science in a Postmodern World* (Harrisburg, PA: Trinity, 1998) ISBN: 1563382555.

Van Till, Howard, *The Fourth Day* (Wm B Eerdmans, 1986) ISBN: 0802801781.

Wildiers, Max, *The Theologian and His Universe: Theology and Cosmology from the Middle Ages to the Present* (Seabury Press) ISBN: 0816405336.

Wiles, Maurice, *God's Action in the World* (Harrisburg, PA: Trinity, 1989) ISBN: 999694861.

Wolterstorff, Nicholas, *Reason within the Bounds of Religion* (Wm B Eerdmans, 1994) ISBN: 0802816045.

CTNS Bibliography

Worthing, Mark W, *God, Creation, and Contemporary Physics* (Minneapolis: Fortress, 1996) ISBN: 080062906X.

Research Level

Austin, James H, *Zen and the Brain: Toward an Understanding of Meditation and Consciousness* (MIT, 1998) ISBN: 0262011646.

Banner, Michael, *Justification of Science and the Rationality of Religious Beliefs.*

Brooke, John, and Cantor, Geoffrey, *Reconstructing Nature: The Engagement of Science and Religion* (Edinburgh: T&T Clark, 1998) ISBN: 0567086003.

Clayton, Philip, *Explanation from Physics to Theology* (Yale University Press, 1989) ISBN: 0300043539.

_____, *God and Contemporary Science* (Edinburgh and Grand Rapids: Edinburgh University Press and Eerdmans, 1997) ISBN: 080284460X.

Devlin, Keith, *Goodbye, Descartes: the End of Logic and the Search for a New Cosmology of the Mind* (John Wiley & Sons, 1997) ISBN: 0471142166.

Drees, Willem B, *Beyond the Big Bang* (Open Court, 1991) ISBN: 0812691172.

Gegersen, Niels, and Van Huyssteen, J, Wentzel, editors, *Rethinking Theology and Science: Six Models for the Current Dialogue* (Grand Rapids: Eerdmans, 1998) ISBN: 0802844642.

CTNS Bibliography

Murphy, Nancey, *Theology in the Age of Scientific Reasoning* (Cornell University Press, 1990) ISBN: 0801481147.

Pannenberg, Wolfhart, *Theology and the Philosophy of Science* (Westminster, 1976) ISBN: 0232512906.

Peters, Ted, editor, *Genetics: Issues of Social Justice* (Pilgrim, 1998) ISBN: 0829812512.

Peters, Ted, *God—the World's Future: Systematic Theology for a Postmodern Era* (Minneapolis: Fortress, 1992) ISBN: 0800625420.

Rae, Murray, Regan, Hilary, and Stenhouse, John, editors, *Science and Theology: Questions at the Interface* (Edinburgh: T&T Clark, 1994) ISBN: 0802808166.

Richardson, W Mark, and Wildman, Wesley J, editors, *Religion and Science: History, Method, and Dialogue* (Routledge, 1996) ISBN: 0415916674.

Russell, Robert John, William R, Stoeger and George V, Coyne, editors, *Physics, Philosophy and Theology* (University of Notre Dame Press, 1988) ISBN: 0268015767.

_____, Nancey Murphy and CJ Isham, editors, *Quantum Cosmology and the Laws of Nature: Scientific Perspectives on Divine Action* (University of Notre Dame Press, 1993) ISBN: 0268039763.

CTNS Bibliography

_____, Nancey Murphy and Arthur Peacocke, editors, *Chaos and Complexity: Scientific Perspectives on Divine Action* (University of Notre Dame Press 1995) ISBN: 0268008124.

_____, Stoeger, William SJ, and Ayala, Francisco, editors, *Evolution and Molecular Biology: Scientific Perspectives on Divine Action* (Vatican City State and Berkeley: Vatican Observatory and CTNS, 1998) ISBN: 0268027536.

_____, Nancey Murphy, Theo C Meyering and Michael A Arbib, editors, *Neuroscience and the Person: Scientific Perspectives on Divine Action* (Vatican City State and Berkeley: Vatican Observatory and CTNS, 1999) ISBN: 0268014906.

Stenmark, Mikael, *Rationality in Science, Religion and Everyday Life: A Critical Evaluation of Four Models of Rationality* (University of Notre Dame Press, 1995) ISBN: 0268016518.

Torrance, Thomas, *Theological Science* (T&T Clark Ltd, 1996) ISBN: 0567085147.

_____, *Reality and Scientific Theology* (Scottish Academic Press, 1985) ISBN: 0707304296

Author Index

Albright, C 199
Alston, P 119
Alszeghi, Z 95n
Ambrose, 22
Aquinas, T 54, 56 - 58, 61
Arbib, A 53n, 92n, 209
Ashworth, W 32, 47
Ashbrook, J 199
Athanasius, 45
Augustine, 7, 22
Austen, J 207
Ayala, F 75n, 209

Babbage, C 36, 37, 45
Bacon, F 10, 11, 32, 33, 45
Baillie, D 185
Balfour, T 45
Banner, M 207
Barbour, I 62n, 195, 201
Barr, J 202
Barry, J 195
Bartholomew, D 186, 202
Baxter, R 34, 45
Bellarmine, C 10
Birch, C 181, 182
Bohr, N 99
Bonaventure, 27n, 45
Bono, J 21n, 33, 195
Borgmann, A 123
Bowler, P 15, 16
Breck, A 121n
Brennan, M 45

Brockelman, P 199
Brooke, J 16, 20n, 180n, 195, 202, 207
Brothers, L 61
Brown, J 159, 165n
Brown, R 174n
Brown, W 53n, 81, 84n, 87, 103n, 113, 202
Browne, T 21, 22, 46
Buckland, W 7
Butler, J 46

Calvin, J 30, 31, 46
Campbell, D 74, 75, 82
Cantor, G 20n, 195, 207
Cantwell-Smith, B 101
Chadbourne, P 38, 39, 46
Chrysostom, J 22, 25, 26, 46
Chung-Jen, T 101
Cicero, M 46
Clayton, A 207
Clayton, P 92, 118n
Clifford, R 176N
Coates, G 176N
Cobb, J 113
Cole-Turner, R 199, 203
Collins, J 176n, 200
Copernicus, 11
Corsi, P 16
Cosslett, T 196
Coyne, G 208
Curtius, E 21n, 27n, 48

Author Index

Dante, 46
Danielson, D 16
Darwin, C 7-12, 16, 38
Davies, P 159, 165n, 199
Dawkins, R 181
Deacon, T 58n, 84, 85
Dennett, D 81n, 85, 86, 181
Descartes, R 3, 32, 54, 55, 91
Desmond, A 16, 17
Devlin, K 207
Dick, T 46
Dillenberger, J 202
Dirac, P 157
Dobbs, B 196
Dobzhansky, T 75n
Dove, J 46
Draper, W 4, 5, 20
Drees, W 207
Dretske, F 72
Durant, J 202

Edelman, G 77n
Edwards, D xii, xiii, 193n, 199
Edwards, P 80n
Ellis, G 200

Fagg, L 203
Fantoli, A 17
Ferngren, G 20n, 196
Ferris, T 200
Fitzmyer, J 74n
Flannagan, O 110
Flood, R 159n
Force, J 17, 196
Frye, R 203

Galilleo, 5, 10, 11, 33, 34
Gassendi, P 32
Glacken, C 196
Gilkey, L 203
Goodenough, U 188
Görman, U 65n
Gould, S 13, 14, 17, 93, 181
Gray, A 6, 7, 12
Gegersen, N 208
Greenspan, R 83n
Greshake, G 97
Gunton, C 111, 141, 142

Hagoort, P 58n
Hahn, R 165n
Hare, R 63
Harrison, P 5, 17, 29, 30, 48, 196
Haught, J 151, 183, 203
Hawking, S 157n, 160, 162, 163
Hebb, D 59
Hefner, P 203
Heil, F 72n, 76n
Heisenberg, M 80n 81, 83, 157
Hess, P 22n
Hoane, J 101
Hodge, C 40, 41, 46
Hofstadter, D 85, 86n
Holton, G 48
Huchingson, J 203
Hume, D 161

Inati, S 59n
Iranaeus, 25, 46

Author Index

Isham, C 119n, 208

Jacob, M 196
Jeeves, M 203
Jitse, M 195

Kant, I 55, 79
Kepler, T 11, 32, 46
Kim, J 70, 71
Klaaren, E 203
Knight, D 196
Kurzweil, R 110, 113
Kyriacou, C 83n

Lactantius, 22
Lagercrantz, H 65n
Lamarck, 15
Laplace 6, 156
Lash, N 67
Le Conte, J 39, 40, 47
Le Doux, J 57n, 60
Leibniz, 6
Lille, A 27
Lindberg, D 4, 17, 20n, 48, 156n, 197, 203, 204
Livingston, D 197
Locke, J 4
Lockwood, M 159n
Loder, J 204
Lyell, C 15

Machen, J 41, 47
Mackay, D 81, 82n, 83n
Malony, H 53n, 92n, 113, 103n

Manuel, F 21n, 34, 35, 47, 197
Marcion 25
Martyr, J 46
Matt, D 200
Matthews, C 204
Maxwell, J 7
Mayr, E 181, 182
McDaniel, J 192, 193
McFague, S 204
McGinn, C 72
McLuhan, M 48
McMullin, E 122
Mele, A 72n, 76n
Meyering, T 53n, 92, 209
Miller, H 8, 9
Misner, C 121
Mivart, G 14
Moltmann, J 151n, 204
Monod, J 181
Moore, A 12, 13
Moore, J 17, 48
Morris, H 38, 47
Murphy, N 53n, 63n 92, 103n, 113, 119n, 202, 204, 208, 209
Murphy, R 174

Neidhardt, W 203
Nesteruk, A 17
Newton, I 4-6, 9, 34, 47, 164
Nozick, R 65
Numbers, R 20n, 48, 156n, 197, 203, 204

Author Index

Olson, R 197
Origen, 25
Osler, M 20n, 48, 195, 197

Paine, I 35, 36, 43, 47
Paley, W 43, 180, 183
Pannenberg, W 67, 151n, 208
Pasteur, L 6
Peacocke, A 17, 62n, 92, 160, 169, 170, 186, 187, 204, 209
Peirce, C 122
Penrose, R 159n, 165-167
Peters, T 151n, 200, 205, 208
Planck, M 157
Plato 54, 91
Polkinghorne, J 118n, 127, 200, 205
Pollard, W 169
Pope John Paul II 93, 94
Pope Pius XII 93, 94
Popkin, R 196
Powell, B 14
Prest, J 198
Prusak, B 97, 98
Puddefoot, J 113

Rae, C 208
Rahner, K 96, 97, 147, 151, 185
Ratzinger, J 91, 92, 97
Raven, C 48
Regan, H 208
Rescher, N 80
Richardson, W 208

Rolston, H 205
Romanes, G 9
Rössler, O 165
Rudwick, M 18, 198
Rupke, N 18
Russell, R, 53n, 92n, 118n, 119n, 208

Sabundus, R 19, 27, 28, 48
Santmire, H 205
Scheiders, S 174n
Schmaus, M 95n
Schrödinger, E 157-159
Schusler, H 166
Scott, A 59n
Searle, J 61, 62
Sedgwick, A 7
Shapiro, B 49
Shortland, M 8
Sina, I 58-61
Sobosan, J 200
Soskice, J 205
Southgate, C 198, 201
Sperry, R 74
Stannard, R 206
Stenhouse, J 208
Stenmark, M 209
Stoeger, W 208, 209
St Victor, H 27
Stewart, I 166, 168
Szubka, T 70n

Teilhard de Chardin, P 151n, 206
Temple, F 41, 42
Templeton, J 201

Author Index

Tertullian, 25, 47
Theissen, G 206
Torrance, T 209
Tracy, T 118, 206
Trask, W 21n

Usher, J 172

Van der Meer, J 195
Van Gulick, R 76n
Van Huyssteen, W 206, 208
Van Till, H 206
Varghese, R 204
Vater, B 177n
Voltaire, 9
Von Hohenheim, T 29, 30

Warren, R 70n
Warwick, K 110, 113
Waters, B 203
Webb, C 28, 49
Webster, C 18

Welch, C 198
Welker, M 127n
Wertheim, M 201
Westermann, C 177n
Westfall, R 34, 39
Whewell, W 4
Whiston, W 5, 9
White, A 20
White, E 38, 47
White, P 174
Wilders, M 206
Wildman, W 208
Wiles, M 206
Wilkins, J 11
Wilson, A 13
Wolterstorff, N 206
Worthing, M 207
Yourgrau, W 121n

Zukav, G 165, 166

Subject Index

allegory: 5
anthropic principle: 137, 148
anthrology
  Christian: 67-68
apologetics: 6-7, 39
Artificial Intelligence: 99-112
astronomy
  Copernicus: 11-12
  Galileo: 11-12
authority
  biblical: 5, 9, 28, 32, 54, 175-179

behaviourism: 106
Big Bang: 130-132,
body soul dualism: 53, 55-68, 69

causation
  bottom-up: 76
  downward: 70-79, 91
  mental: 70-72,
  top-down: 186-187
chaos: 120, 165
cognitive neuroscience: 55-68, 79-80
cosmology: xii, 45, 128-145, 178-179
cosmological evolution: x, 140-142
creatio continua: 117, 139
creatio ex nihilo: 117, 139 153

creation
  as God-given gift: 172
  bible stories of: 173-179
  monothesistic account of: 11
  suffering of: 187-193

Darwinianism: 7-15, 39-40, 77, 180-187
deism: 8, 12, 35-37, 43
determinism:
  biological: 52, 65, 85-86
  environmental: 65, 86
  neurobiological: 52, 65, 69, 89
  scientific: 156-158
  vs indeterminism: 88, 108, 159, 164-168
  vs non-determinism: 102-104, 187
direct divine action: 126-127
divine action: 117-119, 125-127, 138-139, 154-155, 160-162, 168-171
and the laws of nature: 119-123, 154
divine intervention: 154, 160-164
DNA: 57, 75-76

epiphenomenalism: 71
eschatology: 66
essentialism: 3, 4-5
evolutionary biology: x,

Subject Index

xii, 3, 7, 12-14, 53, 93-98, 130-132, 135-137, 146-148, 172-194
  randomness of: 172-173, 179-187

free will: 64-66, 69-90, 91, 185, 189

God,
  as
    absolute future: xii, 149, 151
    communion: 144, 148-150, 183
    Creator: 24, 139, 140, 144-145, 160-162, 174, 184-185, 186, 192
    power of the future: 151-152
    relational: 135, 139, 142-144, 149
    tinkering with the world: 138-139, 148, 153-171, 189
    Trinity: 67, 96, 125, 139, 148, 172, 183
    ultimate goodness: 56
  and
    creation: 128-145, 162-163
    evolution: 124-127, 150
    freedom: 160-162, 164-168
    intentions for humanity: 10
    laws of nature:
    nature: 22
    providence: 160-163
    time: 140, 144
  God's action in history: 118, 124
  God of the gaps: 6, 118, 156, 169
  grace: xiii, 85-86

Hebrew view of the world: 53-54,
historicism: 35-37
human person: 52-68,
  ontological classification: 100-101

Information technology: xii, 99-112

knowledege
  natural: 28-29
  scientific: 9
  theological/religious: 4, 21-23, 30, 32, 41-42

Lamarckian evolutionary theory: 15
langauge and metaphor: 19-20
laws of nature: xii, 117-127, 139-141, 154, 160-162

metaphor in religion/scientific discourse: 19-51

216

Subject Index

metaphysics
  Aristolelian: 4, 11-12, 26, 31-32, 54-55
  from below: 140-143
miracles: 161-163, 169-171
monism:
  dual aspect: 92
  emergentist: 92
  reductive: 106
monotheisitc account of creation: 11

neuro-biological determinism: 52
neuroscience and the soul: 56-61, 72-74
Newtonian science: 5-6, 9, 121

papal infallibility: 4
patrisitc thought: 24-26
physicalist accounts of nature: 55
physicalism
  non reductive: 52-68, 69-90, 91-98, 105
Planck era: 132-133, 137
pneumatology: xii, 125, 144, 151, 183-184, 191, 192-193
predestination: 64, 167-168
prophecy
  biblical: 35

quantum mechanics: 3, 155, 156-159, 166
quantum physics: 159

quantum theory: 120, 157
Qur'an: 23

reductionism: 69-70
resurrection of the body: 53-55, 66, 9-97
revelation: 23, 26, 28, 32, 35-37, 41, 45, 128, 172, 191

science and religion dialogue: *passim* 3-15, 20-45
scientific revolution: 32-35
Scripture:
  Christian: 4-5, 7, 10, 21-23, 26-29, 35-37, 42, 45, 55, 173-179, 189
  Hebrew: 23-24, 42, 55
soul:
  and the neurosciences: 56-61
  and the senses: 58-61
  creation of: 95
  functions of: 57-58
  immortality of: 52
  powers of: 59-61
stoicism: 24-25
supervenience: 62-64

theodicy: 12, 56, 184, 188
theology,
  biblicaly based: xii, 28-29, 53, 173-179
  of creation: 128-145, 173
  medieval: 4, 22, 26-29,

Subject Index

54-55
natural: 21-23, 41
prophetic role of: x-xi
Protestant: 5, 28, 54-55, 97
Roman Catholic: 4, 5, 9-10, 14, 92-93
Reformation: 29-31,

Thomism: 54-55, 58-59, 94
time: 97-98, 137-138

trichotomism: 105
'Two Books' metaphor: 19-44

world-view
 Hebrew: 23, 53-54, 173-179
 Hellenistic: 53